ZEN AND THE RESCUE DOG

JOURNEYING WITH YOUR DOG ON THE PATH TO ENLIGHTENMENT

KJ FALLON

A POST HILL PRESS BOOK

ISBN: 978-1-64293-110-5
ISBN (eBook): 978-1-64293-111-2

Zen and the Rescue Dog:
Journeying with Your Dog on the Path to Enlightenment

Cover art by Cody Corcoran

This is a work of creative nonfiction. While all the stories in this book are true, some names and identifying details have been changed to protect the privacy of the people involved.

This book is not intended as a substitute for the medical advice of physicians or veterinarians. The reader should regularly consult a qualified medical professional in matters relating to his/her health and the health of their dog, particularly with respect to any symptoms that may require diagnosis or medical attention.

Post Hill Press, LLC
New York • Nashville
posthillpress.com

Published in the United States of America

Advance Praise for *Zen and the Rescue Dog*

"Happiness is a warm rescue puppy, and a copy of *Zen and the Rescue Dog*. This wonderful, touching book articulates for so many what has helped bring the end of suffering thanks to the canine souls in their lives. For dog owners, for dog lovers, for students of Buddhism, and for anyone curious about all of the above and a path to happiness, this book is for you."

—William Norwich, Author of
Learning to Drive and *My Mrs. Brown*

"There are many paths to enlightenment. For Robert Pirsig in 1974, the path of Zen famously involved a motorcycle trip with his son, while for KJ Fallon, it's been through the experience of adopting and caring for rescue dogs. Her book, *Zen and the Rescue Dog* offers a beguiling how-to guide for those who would take this path. There's something for every dog lover in this book, and much practical advice. Even for those not interested in embarking on the Eightfold Path, Fallon makes a compelling case that the very acts of adopting and caring for these animals, who cannot speak of their past suffering, brings us out of ourselves and puts us on a path to goodness."

—Eugene Linden, author
of *The Parrot's Lament*

"Buddhist masters tell us that the path to enlightenment requires us to live in the moment, without agonizing over the past or worrying about the future. It can feel like an impossible task, but if you live with a dog, you're living with a highly sentient being that does exactly that. And in the marvelous and insightful book *Zen and the Rescue Dog*, KJ Fallon shows that the deeply interdependent relationships humans have with their canine companions help us to be centered and loving even if we never meet a monk or enter a monastery."

—Michael D. Lemonick, Chief
Opinion Editor at Scientific American
and the author of *The Perpetual Now*

"For many mornings when it was just my dog Topsie and me, we would greet the morning together. All others asleep, my coffee ready, her chin raised to me to be scratched and loved. It was our shared Zen. There are clear reasons the words 'Dog' and 'Zen' have the same number of letters, as does the word 'one,' as being one with your dog. In case you forget why or how to do that, KJ Fallon guides you back, compellingly showing how dogs and Zen meld, as if birthed together. KJ precisely reminds us of the uniqueness of rescue dogs in rescuing humans and masterfully urges us to see the whole dog in this embracing book. Dog and Zen may not be the first thing you think of or think of at all. Now you know thanks to KJ Fallon. And just wait until they place their paw on you. Leave it there and don't move."

—Tom Squitieri, Award-winning for-
eign correspondent, poet, and dog lover

"When life throws us blistering curve balls, many try to deflect or minimize their impact. Others crumble into a dark place. How to cope? *Zen and the Rescue Dog* lights a path for embracing tough moments and keeping focus on the present. Even without a dog of your own, you'll likely find that KJ Fallon's lessons will resonate. And who knows? You just might be persuaded to adopt a canine companion of your own."

—Wendy Cole, Managing
Editor of REALTOR Magazine

"For those who are tired of simply mindful breathing but are not yet ready to embrace the entirety of life as a patience-teaching experience, KJ Fallon reminds us that there are rescue dogs. Fallon's genius insight is that every rescue dog arrives marked by a (usually mysterious) past life. And that this, along with the ongoing and loving diligence of adopting and raising one, is a wonderful vehicle for teaching the dharma. It turns out the Eightfold Path is strewn with kibble."

—David Van Biema, Former lead
religion writer, *TIME* Magazine

"KJ Fallon's *Zen and The Rescue Dog* is an original, insightful, superbly written exploration of how caring for traumatized rescue dogs can heal us even as we restore them to health and keep them happy. Applying principles of Zen to dog care might sound far-fetched but Fallon demonstrates that it is anything but. Dogs can teach us a lot. They live in the present as far as we know. Unlike people, they do not dwell on past pain and suffering or on aspirations that might never be fulfilled. Among other things, Fallon shows how such simple practices as stroking or walking a dog can be a kind of meditation, centering us in the present as well. Dog care can diminish our anger or worry and enhance our well-being. The book is also filled with practical information on caring for rescue dogs. Most dog lovers would find this book of interest. It should be required reading for dog lovers who are preoccupied with past disappointments and anxiety about the future—as so many of us are."

—Dr. Christopher Hanson, Associate
Professor of Journalism, University
of Maryland, College Park

"Practicing Zen Buddhism with your pup may sound barking mad, but I'm grateful to have been shown the Eightfold Path herein. As a Tibetan terrier, our dear old rescue dog seems drawn directly to the Path. Mindful of its benefits to both, and ever thankful to KJ Fallon for guiding us there, Patches and I become Buddhists together again each morning."

—Patrick Tracey, author of
Stalking Irish Madness

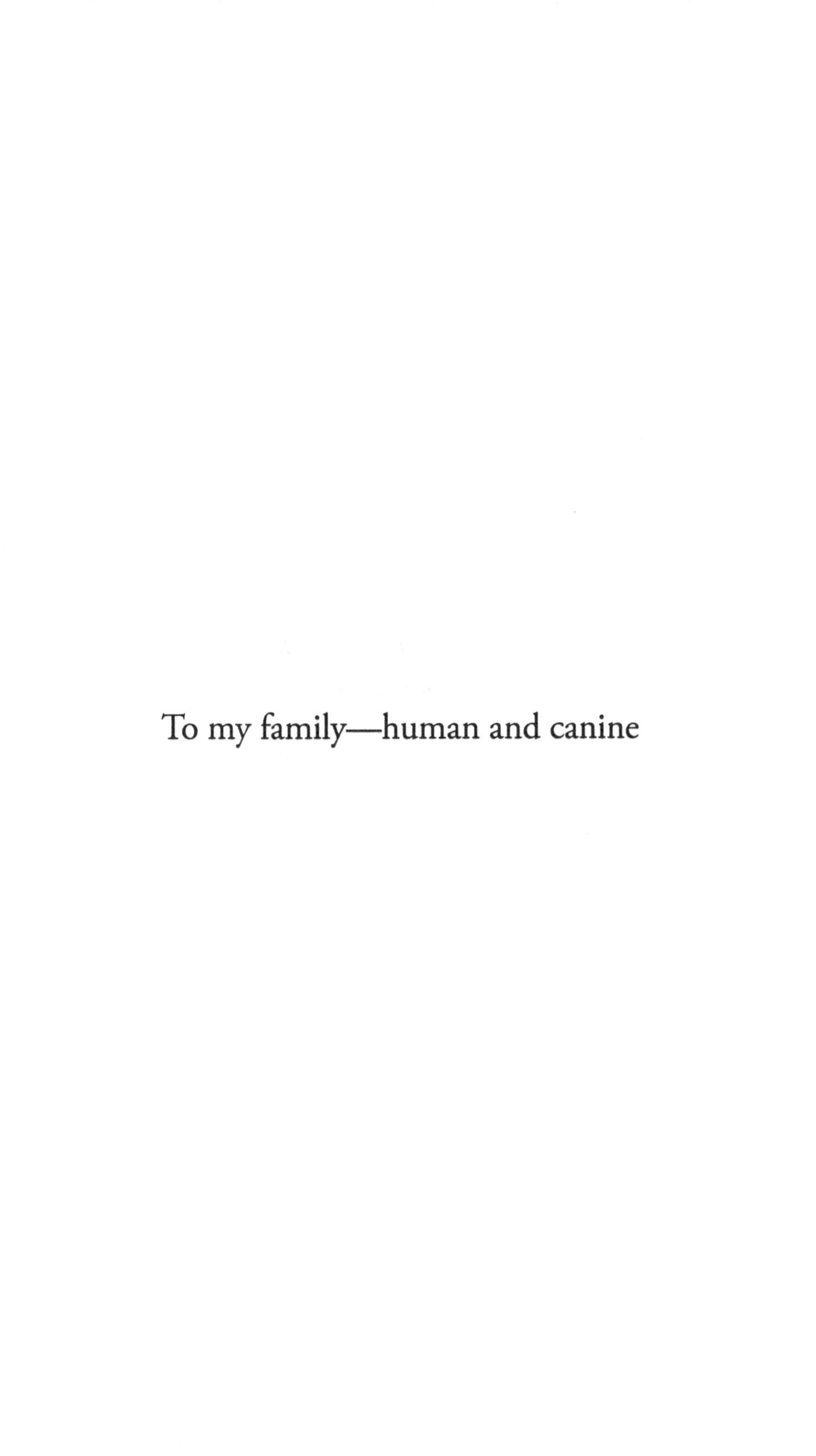

To my family—human and canine

Be kind whenever possible.
It is always possible.

—Dalai Lama

Before enlightenment:
feed the dog, walk the dog.
After enlightenment:
feed the dog, walk the dog.

—Variation on a Zen Buddhist Proverb

CONTENTS

FOREWORD

OF SHAGGY DOGS AND BODHISATTVAS

I knew I was stumbling onto liberating ground the minute I entered KJ's heartfelt and compassionate reflections on how a dog, especially a rescue dog (a symbol of the homeless, the wounded, the neglected), can bring us into a bracing, keen, and kind sense of the here and now. You only have to look at a dog, her eyes watering as she sits beside you at the dinner table, hurries out into the driveway to alert a stranger that you're in desperate need of help inside, waits—sometimes for days and months at a time—by the door for you to return, for you to realize that no one could teach us more about steadfast loyalty, selflessness, and unconditional love.

Yet, when it comes to a rescue dog, you're reminded of something even deeper: many dogs, who have known suffering too well, are the ones who can ease our own suffering and loneliness, our sense of being forgotten. They don't whine or complain, they've left their pasts fully behind, but in their eyes, their reflexes, their sniffing noses, you catch an alertness, often a tail-wagging exuberance, that all of us could begin to learn from. Does a dog have a Buddha nature? Yes, and sometimes it's more visible than that in a human being!

The challenge, for us, is to retrieve a deeply human nature that's worthy of our friend's Buddha nature and able to see the two as one and the same. Be content with the "master" or "mistress" you have, and don't expect the neighbor to offer anything more. Don't lose yourself in what-ifs, but enjoy that bouncing ball and this great stretch of beach right here. Confronted by a textbook on a cold day in the forest, remember that the best way to be saved is by being consumed not with your mind, but with your canines!

As I read my old friend KJ's sneakily profound meditation on suffering, kindness, and freedom from longing, I remembered sitting amidst a handful of other guests in the antechamber to the Dalai Lama's living room in Dharamsala, all of us seated properly and silently amidst the blazing thangkas. Suddenly, a door to the terrace flew open and in romped a buoyant German Shepherd, who proceeded to jump onto the lap of the oldest and most venerable red-robed figure in the room and lick him all over.

It was no surprise to learn that this puppy was the close friend of the Dalai Lama, the Tibetan leader's latest companion, whose energy and instantaneous affection, whose freedom from self-consciousness and formality, were a daily solace to His Holiness.

I thought of the statue, a generations-old favorite meeting place in busy Tokyo, that remembers Hachiko. Hachiko was the legendary Akita who waited for his master—a professor at Tokyo University—for nine years, nine months, and fifteen days, through snow and typhoon, not knowing that his friend had suffered a cerebral hemorrhage and passed away. The patient dog was fed by passersby and local shopkeepers and is revered, decades later, as one of the country's most beloved models of friendship and faithfulness.

And I recall the times when I have been in Tibet or Bhutan or Ladakh, high up in the Himalayas and entering a temple courtyard have been met by a large array of dogs lined up in front of the prayer hall. They sit quietly, most of these mutts, thirty or forty of them at intervals across the sunlit space, and they never make a fuss or get into fights. Some people say they're there for the food or the protection; some say they were monks in a previous life and having been demoted through some misbehavior, are now awaiting their chance to be back amidst the sutras again.

All I can say is that these dogs often act better than we do and embody a watchfulness, a steadiness, a calm that I would be lucky to find within myself. A celebrated story in Asia tells of a young man who was going to see the Buddha and promised to bring his aging mother a keepsake from the teacher. As he was returning home, he remembered—too late—that he'd forgotten to ask the Buddha for a memento.

Chancing to pass a dead dog along the road, he pried out a tooth from the corpse and presented it to his mother as the tooth of the Buddha.

The old woman beamed and placed it on her altar. Very soon her devotion to the tooth, through daily prayer and attention, caused it to pulse and glow, as if it really were a relic of the shining prince. Everything, in short, can be luminous, if only we bring the right eyes to it.

Pico Iyer,
Banff Centre for Arts and Creativity
August 2017

INTRODUCTION

DOGGY STEPS TOWARD ENLIGHTENMENT

This is a book about how, if we let them, dogs we have adopted or have cared for in some way can help us to be better humans. I am an expert when it comes to the dogs in my life or have been a part of my life. Our animals need us to do the talking for them and at times it can be hard to figure out exactly what they want and need. But maybe the problem is on the human side. We need to be able to take the time to really get to know our pets so that we can be there for them and, in turn, let them help us.

Sharing your life with a dog keeps you grounded and in the present. And, just as one thing leads to another in caring for your dog, the path to Enlightenment involves a natural continuation of interconnected steps. One step leads to the next and the last step on the path is really the first step in the continuation of the journey. The step you take today will lead to the step you take tomorrow. These steps are not fixed like the steps in a staircase. They are nonlinear and more fluid. Concentric, but not quite.

Sharing life with a pet who has been rescued takes this a bit further. The pet's person might be aware of the dog's likely sad past or may know next to nothing about how the dog grew up. While the dog is focused on its present circumstances with his or her human, the dog can still be imprinted with what has happened in the past. Abuse leaves a very deep scar. A dog can be hand-shy or reactive to sudden movements made by his human until they each get to know each other. Just as Zen needs to be experienced rather than just learned, so it is with a rescue dog. Of course, learning how to properly take care of any pet is a must, and that includes the human learning from how the dog reacts to certain gestures or objects; in time, the dog will hopefully learn to not fear these gestures or objects. Experience can lead to learning but learning is more external and not integrated if it is not accompanied by experience.

With my work as a journalist, I was constantly focused on what was happening in the news and I was enmeshed in every aspect of current events: business; politics; entertainment; lifestyle; the arts; fashion; design; and more—meeting deadlines; researching; reporting; and writing on a variety of subjects for both fast-breaking and longer, in-depth issue-related topics. I interviewed prominent newsmakers from top financial CEOs to A-list celebrities and pop stars, to a future president of the US. But now it was time to find a more Zen-like alternative. Something that was the antithesis of the work I had been doing for so long.

There was also that imbedded need to want to be simply present—not think and just *be*—with my dog. My journalistic tendencies drove me to want to find out the best way to be able to stop and focus only on my dog, really pay attention only to her for five minutes at first, then for longer periods.

One day, when I was sitting alongside my dog as she was resting quietly, I started gently and slowly stroking her side and kept on in an almost slow, rhythmic movement. There was soft meditative music playing in the background. Suddenly, it occurred to me that what I was doing and feeling while I was letting my hand glide softly and repeatedly along my dog's soft side was a form of meditation. It wasn't long before I realized that using the steps of Zen Buddhism like a kind of mystical malleable ladder would be a good way to align what I was trying to do with how my dog lives and might also help anyone who wants to know their dog better and have a more in-the-moment life like their dog and with their dog. I have had a lifelong interest in Buddhism and the lessons and steps for attaining Enlightenment. And shouldn't we all strive to be bodhisattvas? Think of the Archangel Chameul in some Christian and Jewish traditions. Chamuel's focus is to help us love others in a manner that benefits both the one who loves and the one who is loved. Chamuel inspires us to assess and purify how we think, act, and feel, which in turn leads to peaceful relationships with others and with ourselves.[1] The search for inner quietness and calm crosses many beliefs and philosophies.

Buddha advises, "The secret of health for both mind and body is not to mourn for the past, worry about the future, or anticipate troubles, but to live in the present moment wisely and earnestly." How can we train ourselves to focus on the present and savor what is happening, or not happening, now? When you are dealing with something that can be frustrating and require the utmost patience, like walking your dog and waiting for what seems like forever for her or him to do what you brought them outside to do—late at night, in the rain—it is hard not to get frustrated thinking about what else you could be doing, like sleeping.

I knew how to observe others and listen to their stories and write about them and now it was time to observe and listen to my dog, and learn how to live with her in the present tense. What better Zen Master than my selfless and soulful companion who always lives in the moment?

To the Reader: What I Hope This Book Will Mean to You

It is my goal to show some ways to make peace with the changes and upheavals that may at times turn the path to Enlightenment into a rocky, rut-filled, almost unnavigable road. Not only does there always seem to be something new to be done, some unexpected issue that needs to be addressed, above and beyond what we have set aside time for, but there are the many, sometimes sneaky, distractions we ourselves create.

I have put together my experiences with my several rescue dogs along with some thoughts from other dog adopters, dog experts, and professionals from well-known rescue organizations and followed these with how they relate to the path to Enlightenment. I have also included some ideas that you, the reader, and your dog might find helpful to use as you both journey along your path to Enlightenment. Suggestions include meditation and journaling. Journaling is a centuries-old way to discover more about yourself and try to make sense of all of the ideas and thoughts you carry with you, so it follows that journaling can help you to let your rescue dog guide you on the path to Enlightenment. I have included some journaling exercises and other ideas that can help you with trying to stay centered in the moment and just being with your dog.

If we let our rescued canine companions guide us, the path can become a little smoother, a little less rocky, and a lot more traversable. All we must do is to really pay attention to our pets and see how they are able to just...be.

The essence of Buddhism that we will be following throughout the book:

The Four Noble Truths

1. Life is suffering.
2. The root of suffering is desire.
3. We need to stop the desire.
4. We can end the desires by following the Eightfold Path.

The Eightfold Path

1. The Right View
2. The Right Intention
3. The Right Speech
4. The Right Action
5. The Right Livelihood
6. The Right Effort
7. The Right Mindfulness
8. The Right Concentration

PART ONE: THE FOUR NOBLE TRUTHS OF LIFE WITH AN ADOPTED DOG

CHAPTER 1

THE FIRST NOBLE TRUTH: LIFE IS SUFFERING

We go about our days thinking about what we need to be doing next. Not thinking about what we are doing at that moment becomes a habit and when serious difficulties suddenly arise—losing your job, finding out you have a serious illness, or that a loved one (human or canine) is seriously ill—we don't want to think about that at all. Except maybe in terms of "How could this have happened?" or "How did I not know this might be in the future?" or "If I only had done this or that..." The idea that something bad could have happened when we weren't looking is unacceptable. We don't see that we really don't have any control anyway. Not when it comes to the future.

Sharing your life with a dog can guide you on the path to Enlightenment—the road to becoming more aware of and content with what you have and where you are at this moment. Animals naturally live in the present. Humans, on the other hand, seem to be either focused on the past or obsessing about tomorrow.

Several years ago, I adopted a young black mixed-breed dog and named her Charlotte. She has taught me a lot about patience, staying focused, and living in the present. She is very serious about making sure I am giving her my attention when she needs it—and when she demands it. After living with me for only a week or so, she learned that if I came into a room where she was and I was carrying a laptop or phone, it meant I was not going to be giving her my undivided attention; she barked in protest until I set the device aside and focused on what she wanted to, often playing with a squeaky toy.

Charlotte has helped me to accept that even though she has suffered, her suffering has not defined her. She had and still has some serious issues, and she has taught me not to freeze in fear when she is so ill that she has to be brought to the animal emergency hospital. She also has some phobias that I can find no reason for, and yet I must understand them and help her through her fears. I have learned to accept that for her, suffering has happened and might happen again, and we both just have to live through it. I have to do what I can for her.

* * *

To begin at the beginning is impossible. I don't know what the beginning was for sweet Charlotte, my rescue pup. She is a "big little dog." She has the temperament and personality of a little dog packed in the body of a somewhat bigger dog. Miniature poodle/Wheaton terrier attitude in a pharaoh hound/Belgian shepherd. Her life in a before we adopted each other can only be known from sparse records and comments along with what her foster mom had to say. Evidently, Charlotte was cat-like during her two-week stay in foster care, where she lived with a cat and tolerated frequent visits by a much bigger dog. It would seem then that Charlotte might be comfortable around other dogs, even dogs much bigger than she is. That is another story and time would tell that tale.

I can speak only to the beginning of us, of how my life with her began and how her life began with me. We knew from the picture of her as a puppy on the rescue website that she would be the one for me and my family. There was a depth and poignancy in her eyes, as if you could see deep into her soul. She seemed an old soul, not unlike our rescue dog Alice, whom we had recently lost.

Life had to have been harsh for my dog before we found each other. My family and I picked her up at the home of the woman who was in charge of the animal rescue organization that had put Charlotte up for adoption on their site and where her foster mom was going to drop her off. While we were waiting, she told us how it was great I was adopting Charlotte since so many people overlooked black puppies and black older dogs.

Soon Charlotte made her grand entrance with her foster mom and was running around loose, looking for scraps to eat, begging for anyone to give her whatever food was being prepared in the active kitchen and dining area. She looked bigger than she had in her photo. We tentatively approached each other and then I attached a thin leash to her collar while listening to the instructions about her diet, what shots she would need, when to have her spayed, and so on.

Charlotte walked with us to the car and I picked her up and held her during the ride home. It is always fascinating when you first bring an adopted animal into your environment. He or she is completely unfamiliar with the surroundings that you live in and know. She slowly and falteringly started exploring the entryway, venturing a little bit farther into the room one pawfall at a time. It helps if you bring a few things they have become used to wherever they were before you adopted them, such as a bed, toy, or blanket. The feel and smell can comfort them.

Charlotte was tired from her journey and the new experience. She went to her little bed that her foster mom had given us along with a few stuffed animals and a blanket. To say she was shy would be an understatement. There, she stayed for a long time, sleeping, curled in a ball so tightly wound that there was no prying her to relax. She was a complete furry circle. Gradually, she realized she was hungry and she looked at me to see what was next. I gave her the same food she had been getting in foster care and while she eagerly devoured it, it didn't seem to be the best food for her, given the later aftereffects. Finding the right food for her would take some time and trial and error, and that was okay. She was and is always eager to eat anything.

At first Charlotte was only comfortable being held if she was wrapped in a blanket. As she became more at home and we interacted more, I could examine her more closely. She had a mysterious and ominous scar on her neck. Her deep black fur highlighted the white pharaoh hound streak of scarred skin. She was very nippy, and anyone who has ever raised a puppy (especially a lab mix, although probably every breed and mix has the urge to gnaw, nip and chew while teething) knows all about how the sounds of "Aw," and "Ooh," "He/she's so sweet!" are quickly followed by "Ow!" and "Ouch!" and of course, the often-futile but necessary "No!"

The best time to cuddle a young puppy seemed to be right when they woke up from a nap for then you could pet them when they were in a semi-sleepy state before they realized they needed to gnaw on something. This was always the sweet spot of the morning.

According to her paperwork, Charlotte was rescued in Houston, Texas and taken to an animal hospital. She didn't spend any time in an animal shelter. She had to have lived elsewhere before that, possibly overseas. It took many steps to get to know each other and to help her emerge from the shell of her past. For instance, she was very afraid of ordinary paper coffee cups at first. What had happened to her to cause this reaction? Not knowing her history of even where she was before she was taken to an animal hospital in the southwest fueled images of all kinds of terrible scenarios. Had someone thrown objects like a cup at her? Had this been a part of some abuse inflicted on her when she was very young? I tried to carefully and slowly reassure her that the presence of a cup meant only that someone was having a beverage and that the cup was just a container that would be safely put away when the user was finished drinking. This took many weeks before she would not cringe and try to flee when anyone had a cup in their hand.

She was also afraid of children and would pull to go back home if we heard them even in the distance. This usually means a dog was not around children when he or she was a puppy.

Then there was that inch-long scar on her neck. What had been the cause of that? Trying to figure what may have been the cause of that brought to mind all kinds of unthinkable scenarios—could someone have tried to cut her throat when she a very young puppy? Had she cut her throat some other way, such as climbing through or over a wire fence to escape some unbearable captivity?

There was also a later experience with fireworks. Most dogs don't like fireworks and with good reason. Fireworks can damage a good dog faster than pretty much anything. Fireworks, unlike thunder, which many dogs also fear, are an unnatural sound. After their sense of smell, a dog's hearing is their most acute sense. Humans hear sound frequency in the range of twenty to twenty thousand hertz while dogs hear sounds in a range of forty to sixty thousand hertz. This means that dogs have an extremely difficult time when they hear very loud noises. Just because a noise is okay with your ears doesn't mean your dog won't be affected.

Charlotte has enough little, feisty dog in her that unless the thunder was particularly loud and close, she would run up to the door and bark at it—at least when she was younger. Same with fireworks, usually. But what happened more recently with fireworks was unnerving and led to resurfacing questions about her background.

This time, the fireworks were much more numerous, louder, and closer than usual to where Charlotte lives, so the sounds were much more intense than they had been in the past. The racket continued and got louder as an increasing number of fireworks were being set off. It seemed like what had to be the finale was a just a warm-up to even more ear-splitting fireworks. It was as if it would never end. The windows shook slightly, then suddenly, Charlotte became still and stood as if she was frozen. This was unlike her. She stood motionless for about five minutes. It was clear she was having a PTSD episode.

When new behaviors and reactions happen with a rescue dog, it raises more questions about the dog's past. We had always wondered and conjectured about Charlotte's life before she was rescued. Having a DNA test provided some answers (more on this later). But this reaction of being completely still and in a trance-like state was something new and more than a little frightening. Had she been rescued from a war zone? The information from her DNA test about what breeds were in her makeup indicated that this might have been the case. Most of the breeds in her makeup were not identifiable in the US, indicating that she didn't start her life in the US.

She seemed to have so many other phobias. What was behind those? Charlotte was terrified of doorways. Going from one room to another was a challenge and eventually she was able to do it but with some doorways she would—and still does—walk backward through them. She hesitates, turns around, and backs into the next room. If she is in a hurry, she just backs up faster. Even if her body is almost 100 percent through the doorway in question and she is basically inside of the room that the doorway leads to, she will either turn around and leave to enter the room, the kitchen, from another doorway or she will turn around and start backing up into the kitchen from that dreaded doorway. And when she is eating her meals, for some reason she will quickly back out of the kitchen from one doorway in between the portions and come back around through the other door to eat the next portion. She has to have three meals a day in increments; if she is given the whole meal at once, she will choke and likely throw it all up soon after. And her stomach cannot be empty for too long, or she will vomit bile.

When you adopt a rescue dog, you know that you are starting from the point of not knowing much about the dog. A vet told me not long ago that the mindset of people who adopt rescue dogs is different from people who buy purebred dogs. As she put it, purebred dogs might be appealing to some because they have control when it comes to certain traits as far as what the dog will be like, and to some degree this is true. An adopted dog is like a grab-bag present, you never know what you are going to get.

One rescue dog pet parent told me that while she and her family love their rescue dogs dearly, they come with so many unknown triggers and an unknown past and at times it can be very anxiety-provoking to not know much about them. One rescue dog they adopted when he was seven months old proved to be especially challenging. They have always done training with all of their dogs, but this fellow proved to be very aggressive and would start barking uncontrollably; it was hard to pinpoint what would set off this behavior. We can often never really know what caused a rescue dog the suffering and trauma that lead to issues they reveal as we get to know them.

This leads us to the First Noble Truth.

Life Is Suffering

The First Noble Truth is the Truth of Suffering, *Dukkha*. Life Is Suffering. Does this sound overly pessimistic and full of despair? Does this mean that there is always suffering? Think about it. All this is saying to us is simply that there is suffering in life. And there is, as any one of us knows. After all, living means changing, growing, and evolving; there is always some pain that comes with that.

Remembering that there is suffering when you are looking for a shelter dog to bring into your family will help free you to look beyond the pain the shelter pups have endured in the past and see how they can live with you in the present.

While there are many types of suffering, old age, illness, and death are foremost in many people's minds. Pain varies in degrees in intensity and how long it endures, but it is pain nonetheless. Physical pain can be forgotten after a wound has healed. The pain of losing a loved one can linger in the shadows for years and sometimes never dissipate entirely. The loss is somehow always there. It may ebb and flow in intensity over time but it never disappears.

There are different types of suffering and there are differences in how humans and dogs relate to it. And suffering is not just physical pain or anguish. For humans, it can be a dissatisfaction that never ceases; the never-ending search for something else. Your rescue dog is thankful for what she has now, in this moment. While she doesn't think about the future, she can hope for what might come next. Charlotte might be hopeful that when she sees someone going near her toy bin that it means she is going to have play time and when someone brings her food dish to the kitchen counter it means she is going to be fed soon, as does your dog, more than likely. But at the same time, she is content to be where she is right now, at this moment. Hope is not the same as an anxious worrying about the future. It is just that: Hope.

At the same time, she doesn't dwell on the possible abuse in her past when she was a young puppy. Of course, what happened to her has stamped itself on how she reacts to certain things, even though with time this has diminished, except for the recent incident related to the overwhelmingly loud and long fireworks episode. Whatever happened to her is still a part of her.

You are what you know and have known. We will never know what our adopted dogs knew as very young puppies before we found each other.

Here is where humans and dogs part ways on the path of life. Dogs are in the moment, while humans, on the other hand, think about what they once had and no longer have; what they lost. They think about what they want to have. An adopted dog with an unknown past can be happy with himself in your care.

Not accepting that there is suffering results is the greatest suffering. Fighting against the acceptance of suffering only reinforces its influence. So, accept that there is suffering, and you will diminish its power.

And when we accept this, we can take the first step on our journey to Enlightenment. Old age, sickness, and death. These are inevitable sufferings and we cannot escape them. Can we learn to be at peace with them? It was not easy to accept that Charlotte would likely have scary episodes that would cause her suffering and her family would also be suffering because of the fear of losing her. And I know, as does every reader who loves their dog, that someday, my beloved companion will become old and/or ill, and die. Every so often, a stinging reminder of this emerges.

Realizing there will always be suffering in life will help us to get past that suffering. Our dogs can teach us much about this and can help us to go beyond the suffering in life and realize that suffering does not have to define who we are, just as it does not really define who the rescue dog can become when you adopt him or her.

Our dogs can help us through suffering through their being able to become our companions despite any suffering they have encountered before we knew them. And we must make caring for them a part of our life. If I had known all of the medical and emotional issues Charlotte had before I had adopted her, I would still have chosen to adopt her. Suffering exists everywhere. It just may not be obvious right away.

One way to stop denying that there is some suffering is to look at your dog and find out what can little things you can do to help him or her overcome the suffering that may have formed them in the past.

If you adopted a puppy or older dog that you discovered had some difficult issue, try to find out more about the background of the pup. It is often the case that what you already know is all you will be able to find out. If the dog was fostered for a length of time, often the foster parent can give some clues as to what might be behind your dog's issues. It may be the dog has an allergy to a particular food or treat, or that the dog really is afraid of some common household item, or he or she needs the comfort found in a particular routine.

If you can find out the cause of the suffering behind your adopted dog's medical or behavioral issues, you can better help him or her. But know that you might not be able to. Even if you cannot find a shortcut to understanding what is behind the difficulty you are having with her, you can quietly observe your dog when she awakes in the morning. Look at her—or him—closely and know that whatever trauma has made them who they were when you adopted them, you and she—or you and he—can work together to make them more comfortable being a part of your family.

You had no control over how he or she came into this world, and you can do your best to care for her right now. But as for what could happen that is beyond your control, relax, take a breath and stay where you are in time right now. Can you accept that suffering is a part of living?

"Rather than being your thoughts and emotions,
be the awareness behind them."
—Eckhart Tolle

Thinking too much—it happens to all of us. And it can get in the way of just being. Even when you are trying hard not to think, you are thinking. You can use meditation to redirect your consciousness so that the many scattered thoughts in your mind, including

"If only I had done this," or "What will happen?" or "Why did this happen to me...to her...to him?"—can give way to letting you open the door to your innermost thoughts, sensitivities, and insights.

Meditation has been proven to benefit humans on many levels, from the physical to the metaphysical. People can meditate in many ways. Simply being next to your dog while she is resting, feeling her breathing, and noticing the many tiny ongoing movements in quietness can lead to you being completely in the moment and just being along with her. Depending upon whom you ask, meditation is "... continued or extended thought; reflection; contemplation"[2] or a "... means of transforming the mind."[3] Either way, focusing on your rescue dog no matter what he or she is doing at the time paves the way to the next step, the Eightfold Path.

Here is a simple form of meditation to start that can help you to still your scattered thoughts. When you are meditating, you are consciously trying to separate yourself from your thoughts and feelings and you use specific methods that foster this state. Even just sitting or kneeling alongside your dog is a way to mediate that can benefit you both.

In the morning when you usually first see your dog, take some time to really look at him or her. (If he needs to go outside to take care of his morning business first, see to that and then bring him or her back inside or find a quiet place outside.) If your dog is sitting or lying down, gently stroke his fur in slow movements. Keep repeating this motion and you may soon find yourself and your dog in a trance-like state. And it is a beautiful place to be. Do this for as long as you can, or for as long as your dog lets you. Don't think of anything other than the feeling of your dog's fur under the palm of your hand as you gently move your hand over his head, down his back; feel his feet and notice the texture. Feel his legs and sense their strength even in repose. There is nothing else in these moments, just you and your dog and your sense of his breathing and his sometimes lightly twitching body. This is a form of meditating that you can do at any time of the day or night when you are with your dog in a quiet place. It is also a good way to tell if anything is developing on his or her skin that shouldn't be there, like small cysts.

Afterward, write down in a journal to note how you felt before you meditated with your dog and how you feel now. Do you feel calmer now and readier to allow yourself to go slower and take the time to breathe in the morning, accept the new day, and let go of the events of the time before now?

Perhaps the most emotionally debilitating suffering is the suffering caused by desire. And that is a story for the next chapter.

CHAPTER 2

THE SECOND NOBLE TRUTH: SUFFERING IS CAUSED BY DESIRE

Let's say you do realize that there is a lot that can happen in life that you cannot control, no matter hard you will it. You might want to ensure that you have a pain-free life and that your adopted dog is always protected from discomfort, but that's never going to happen. Accepting that suffering is part of growth and change, and not wishing it were otherwise, is living in the current moment, whatever that present moment entails.

With your adopted dog, you can't always get what you want. Or what you think you want. I remember one time seeing a woman struggling to carry her very big dog, some kind of retriever, from the parking lot to the vet's office, which was a pretty good distance especially if you are carrying a dog that weighs at least eighty pounds. It was clear that her dog must have been terrified of going to the vet's and this was the only way she could get him from the car to the vet office. We do what we must for our pets.

"I am always amazed if I meet a pet for an initial post-adoption exam who has significant health or behavioral issues," said a vet I spoke to. "The idea of returning a dog to the situation it came from is absolutely not an option for most owners and rarely do animals get returned. The need to be needed is very strong and very endearing." Sometimes mismatches can occur, though. "People may not fully appreciate what they are getting into," she added, "like a high energy dog for an elderly sedentary couple, or a dog or cat with fearful tendencies going into a busy household with lots of children. But, even these situations tend to work out somehow."

During her first visit to a vet, Charlotte was so anxious about being there that she made herself flat on the floor, so there was no getting her to walk on her own in any fashion through any doorways into the exam room.

Charlotte's fear of doorways was not that much of a problem at home since she would always find a way to navigate from one room to another. She would adapt her walking style and the direction she faced and somehow get from one location in the house to another. Since doorways were involved at the vet's office, picking her up was the only way to get her to the next step. Once inside the exam room, the vet techs looked over the scant information there was in her meager medical chart from a pet clinic in Texas, where she had arrived from no one really knows where. Later, we would find out more about Charlotte's heritage and her likely origins, but at that moment, she was a jittery little riddle wrapped in an insecure mystery inside of a very energetic enigma. Whatever happened to her is still a part of her. Helping Charlotte to leave her past behind would help her be more comfortable and move forward more freely.

When I adopted Charlotte, she was more than six months old as it turned out from the vet's examination of her teeth. This meant that whatever traumatic life she had before she was adopted, she had experienced it for longer than first thought. The vet also discovered that Charlotte had mange. There were little bald patches that at first were not that noticeable, but a close examination of Charlotte's fur and skin at the vet's revealed that she did indeed have demodectic mange. No wonder she was a tight ball unto herself that would open up occasionally and for food. Food would always entice her. She did, and does, have a very hearty appetite.

My previous dog Alice was always a picky eater, and I had to supplement her regular food—which she often left untouched or barely eaten—with nutritional alternatives that were tasty, like the one that comes in a tube and gives the pup the extra calories they need.

Charlotte, in contrast, who loves to eat and will eat anything, never had a problem having an appetite. So, we reason that if Charlotte ever turns away from food, it will mean that something is seriously wrong with her health. But we would never ever want to let things get that far.

We didn't know much about mange, and we watched while the vet scraped Charlotte's skin to procure a specimen to examine under the microscope to determine if what was causing her bald patches was indeed mange. The procedure was obviously painful for Charlotte. The creepy crawling mites were visible under the microscope and a diagnosis of demodectic mange was confirmed. The next step would be a series of shampoo treatments with a special mange treatment shampoo, which would be done at the vet's.

Here is something we discovered when Charlotte had mange and we were trying rid her of that: the mange shampoo could go only so far. We did not want to subject her to any more scrapings to find out if it was actually mange, which we already knew it was. As a tick and flea repellent, we use a natural topical blend of oils—peppermint, cinnamon, lemongrass, clove, and thyme oils on Charlotte. Another family member had read that this repellent helped to prevent fleas because the oils keep the fleas from communicating with each other. It attacks the insects' neurotransmitter—called octopamine—which is only found in insects such as fleas, ticks and mosquitoes.

So, it might follow that the oils would work on eradicating the mites causing the demodectic mange. We also regularly used a natural flea and tick wipe for daily use on Charlotte during high tick season and when the temperatures were warm enough for ticks during other times of the year. These wipes contained lemongrass and cinnamon oils. We used them around Charlotte's neck, the area that the topical drops would not reach, since the drops are applied down her back from the base of her neck to the base of her tail. We also discovered that the area around her neck where we applied these wipes was starting to become mange-free. So, would using the natural flea and tick repellent work even better?

We applied some of the oils to the area where the mange was present. After a few applications, the mange started to disappear. With faithful application of the oils to any remaining mange areas, it was not long before Charlotte was mange-free.

Charlotte has very delicate features and her chin and nose are very small. During one of the mange baths at the vet's, she somehow got some water inside her lungs, developed pneumonia, and had to go on two rounds of antibiotics. Clearly, this little girl had a fragile constitution and needed to be handled with the utmost care.

Even her spay surgery did not go smoothly. Since the hospital did not have a technician or any staff member there overnight, I picked up Charlotte the evening of her surgery after she was safely brought around from the anesthesia.

About the anesthesia, I have learned that knowing your pet's genetic makeup can help when it comes to medical care and what to be on the lookout for and to try to prevent in the way of possible future health issues. In Charlotte's case, we found out she was part miniature poodle, part pharaoh hound, part Belgian shepherd, part Wheaton terrier plus some unknown breeds. With hounds, there is a concern about anesthesia. Charlotte had enough pharaoh hound in her so that the type of anesthesia being used was of genuine concern.

A few days after her surgery, another family member noticed that the scar from the operation didn't look right and that she might have developed an infection at the surgical site. We brought her back to the vet and the doctor was very reluctant to give her an antibiotic, but at our insistence, the antibiotic was given. After a few days, Charlotte's surgical wound looked much better. We knew we would be taking her to a different vet.

When thinking about what you want for your adopted dog, it is better to take the long view about what is really best for him or her. Take the matter of getting your dog used to being in a crate. While dogs usually end up feeling safe and secure in a crate in their home, if he is afraid of being confined to his crate for reasons you have yet to discover, is it right that you should force him to go into his crate because it is more convenient for you? There can be many reasons why a dog might be afraid of going into his crate. It is our responsibility to figure out the reasons behind this fear and then to try to work to help the dog overcome it. It may be that the dog was confined to a small space for long periods of time at one point in her life.

The crate fear could be compounded by separation anxiety. You can try to ease the dog's anxiety with treats to see if this will get her interested in going into the crate. You can also help by putting the crate in an area where there is lots of activity so the dog doesn't feel isolated. It is also a good idea once the dog gets comfortable going into the crate not to keep her confined for long periods of time. In time, the dog should feel safe in his crate and at home going into it. Even after the dog is comfortable going into and staying in the crate, it is best to limit the crate confinement for a few hours or so at a time, unless it is overnight when the dog is sleeping and you are in the house as well. If you have to be out of the house or apartment for long periods of time, consider alternatives like having a dog walker or pet sitter come in or taking the dog to a reliable doggy day care.

Obviously, there is going to be a period of adjustment for both the dog and the human. The point is that you find a way to make living with you work for both the dog and for you. There are options like a reputable doggie day care—some even have cameras so you can watch your pet on your laptop or phone while you are at work. Make sure there that the ratio of dog to human is going to allow your dog to get the proper attention. Free roaming, as opposed to having the dog being cooped up in a crate all day, is always better.

Back to crate training—you have to work to earn a living, but you want the dog to be safe and comfortable so that at the end of your work day, you both will be happy to see each other but will also have had a positive experience apart. However, you need to remember that your dog has different needs and his own time frame for being ready to do what you want him to do.

I asked a veterinarian about ways to make the transition for the human and the adopted dog to go more smoothly. The vet said it is important to have realistic expectations and not to expect an adopted dog to handle being kenneled very well. If traveling without your pet is a big part of your life, reconsider adopting unless you have a good surrogate family for your pet. Some dogs really have trouble with being left alone if everyone in the household works full time jobs. "Make sure your dog has enough socialization with other dogs and adequate mental stimulation if you need to be away and if he must be boarded," she said. "Doggy day care is a wonderful way to make the day go by for such dogs."

Many dogs need "jobs," and do very well with obedience, agility, and the like. Trainers, behaviorists, and veterinarians can be good resources for a variety of issues relating to fear, anxiety and aggression.

There is no forcing your dog to do what is convenient for you. What works for your dog at the moment may not coincide or even intersect with your needs, so you must work with your dog and understand the "why" of his behavior. With patience, empathy, and a bit of cleverness on your part, in time you will both reach some sort of mutual agreement.

Suffering Is Caused by Desire

The Second Noble Truth is the truth of the source of suffering, *Samudāya*. The root of all suffering is desire, *tanhā*. There's a bit more to it than that—there is seemingly an infinity of steps within each step—but the heart of it is that we suffer because we want.

Accepting suffering is a way to work through suffering. When we are able to accept that there are difficulties and anguish, we can then avoid the endless and fruitless battle to prevent suffering, whether it be in the form of aging, illness, or death. Holding this close to our hearts as far as our pets go also enables us to more enjoy them as they are now. The root cause of the suffering is desire, a desire for things that we think will quench our desire but instead only feeds it.

Pain is inevitable, suffering is optional

But suffering is not just physical pain or mental anguish. For humans, it can be a dissatisfaction that never ceases, the constant of wanting more. Here is where humans and dogs part ways on the path of life. Your dog is thankful for what she or he has now, in this moment. Humans, on the other hand, think about what they once had and no longer have: what they lost. They think about what they want to have. And when they get it, it is never enough. The wanting more never stops.

There are different types of suffering, and there are differences in how humans and dogs relate to suffering. On the surface, wanting what we really don't need can cause not only discontent but a lot of debt. We humans are bombarded with shiny new products; must-have electronic gadgets (that must be constantly updated), places to visit, new fashions to buy, and so much more that makes it hard to just *be*. Just to be okay with, even happy with, what we have already. If you have the latest phone and a newer one comes out, you must have the latest. Now you have the latest but pretty soon yet another new model of the phone comes out and you absolutely cannot live without that and so on, all the way up to whatever is next after that.

Sharing your life with a dog can guide you on the path to Enlightenment and help you to simply enjoy what you have and a large part of that is enjoying being with your dog. Your dog is aware of what is happening at the moment. While how they act and react is often a result of the experiences they have known since birth, it is their nature to be present with what is in front of them or around them now. If your dog seems a little sad because their favorite human or canine friend isn't there, then he or she can be comforted and shake off the sadness if you present him with his favorite toy and then play with him. Or you can offer her a favorite bone to chew.

When it comes to illness and injury, dogs usually do not show their suffering. This is an inherited feature that enables them to hide their injuries from other members of the pack. So, it is not always easy to determine if your dog is suffering from a physical problem and how to determine what it is. This doesn't mean they don't feel the physical pain, only that they will hide it to get on with what they must do. In a way, this is a form of them accepting the suffering. It is our job as their human, of course, to pinpoint the cause and to alleviate the suffering. Humans, on the other hand, like to pretend either that there is no suffering or that it is to be avoided at all costs.

But we can learn from how our pets handle a difficult and painful past. Think of your dog who has been rescued as an adult dog or as a puppy. How are they able to move past their past and how are they able to wag their tail and be happy to see you even when they are not feeling so great? In some cases, the dog may have an especially difficult time, depending on what they had to endure in the earlier part of their life. But somewhere out there is the right match even for the most "difficult" dog. And even dogs who have special needs and require more time, effort, and patience to fulfill even their most basic needs can find a home that is just right for them.

Dogs who have been rescued can have a natural affinity for people who are overcoming obstacles. And here is how they find work so well-suited to them. There are organizations that work to bring prisoners and rescued dogs together so that both can benefit. Sherry Woodard, Best Friend's resident animal behavioral consultant, knows quite a bit about how rescue dogs work with prison programs so that both the dog and the human can benefit. In an interview for this book[4] , she said she works with multiple prison programs, usually in tandem with the head trainers in each of those programs.

The head trainers in turn work with the shelters and rescues that the dogs come from. Then they're working with the prisons and oversee the programs. "They're actually in the prisons making sure that the prisoners are doing kind training," said Woodard. The trainers ensure that the prisoners are using tools and techniques that are helping to improve their skills, which is a very important part, of course. "The people are in prison because they didn't have great skills," she said, "and, also…what happens out of the prison programs is that the extended family has a chance to improve their skills as well." Not only the prisoners, but the extended family of the prisoners, learn from these programs. Often, the dogs are adopted by extended family, or by family friends.

Woodard has high praise for the prison programs, but she realizes that some people are uncomfortable with the idea of the prisoners receiving too much enrichment and having too much fun. But Woodard thinks we should want to them to improve. "I don't think that punishment alone ever works," she explained. "I believe that we should be helping them to improve their skills. Even if they're never getting out of prison, we want them to be better people."

There are a quite a few organizations in different states—Missouri, Kentucky, and Utah, to name a few—that work with rescue dogs and the prison population. Sometimes dogs can get that second chance even before they are turned into a shelter. In Bozeman, Montana, according to Woodard, they have switched from having dogs come in from shelters to having dogs come in from homes where the people have been ready to relinquish the dogs to shelters. The prison program is run similarly to a board and train. The prisoners take in dogs with behavior challenges and work with them, then the dogs go home back to their people.

Rescue dogs being cared for and trained by prisoners is an example of the reciprocal help they give each other. They teach each other. One of the most important things the rescue dog teaches is being okay with things as they are at the moment.

And so, it goes that your dog wants just what he or she needs. Your dog doesn't have to have that fancy jewel-encrusted collar or designer coat. Your dog needs a collar for leash walking and for her tag and she will welcome, of course, a covering that keeps her warm or dry, but she isn't aware of the fancy fabric. Your dog wants food when hunger strikes; to be able to relieve himself when he has to go, and to be able to occasionally attempt to chase that chipmunk that is constantly popping up at the door.

Of course, there has to be some thinking about the future, from what to make for dinner to what time to make an appointment. How do you know when thinking about the future means you are not living in the moment, though? The thinking about the future that is a combustible mixture of anxiety and apprehension is what leads to frustration. It is a wanting that even if the objective is fulfilled, it will never lead to satisfaction. It is like drinking salty water. Your thirst can never be quenched.

How your dog can help you to recognize your unquenchable desires

Take a look at your dog over there. Does he care if you don't have the latest version of the iPhone or if your car is older than your neighbor's? An adopted dog with an unknown past can be happy with himself in your care. Take a cue from the dog and learn to forget about *things* and what you can't change. Try to live more in the present. One way to do this is to be more involved with the dog, not just with his or her basic care.

If you are walking the dog, instead of thinking about what you have to do the rest of the day, or worse, constantly checking your phone, stay focused on the dog and how he is walking. What is he aware of? This is a smart idea anyway since you will be able to respond quickly if the dog reacts unexpectedly to something like a cat or rabbit suddenly darting near or across your path.

Notice how he lifts his head to sniff the air. What is he smelling? What far-off or nearby presence is he detecting? This is the canine's sense of the air, if you will. A dog's sense of smell is about one thousand times more acute than is a human's. And this is a conservative estimate by many accounts. The dog's nose is her most valuable sense. For the dog, the world is in her nose and the nose knows.

Your dog's sense of smell is her introduction to the world around her. Your dog uses her sense of smell to "see" her world. When humans use their sense of smell, we breathe in with the same nasal passage that we use to breathe out the air, so some of that smell dissipates when we exhale. But a dog has two distinct air channels, and a dog uses one just for smelling and the other just for breathing. This means that the dog can store the smell in her nose even while they continue to breathe.

Try to imagine what your dog is "seeing" with her nose. Are you missing something? Are you not fully aware of your surroundings but instead thinking about or looking for (on your phone again) the next pair of shoes you want? It is so easy for our minds to wander instead of focusing tightly on what we are doing.

If you can, find a place to sit or otherwise be still with your dog and really look to see what she finds so interesting. Here is where your phone can help—take a picture of your dog as she is searching the smells with her nose. You probably have a lot of pictures of your dog in your phone, but focus on your dog and also the direction of what your dog is seeing with her nose and her eyes. Then try to find for yourself what is holding her interest. If you brought your journal along, write down what you see and how it feels to be able to experience, as best you can as a human, what your dog is experiencing in that moment. Then write down how being more with your dog makes you feel. Hopefully, you will be more aware and more content. For the time being, at least.

Amazingly, a dog can smell from each nostril separately from the other. This allows the dog to figure where the smell is coming from. The dog's nasal passages have very concentrated olfactory receptor cells. There is also a very good reason for your dog's nose being wet. The mucus on the dog's nose actually helps the dog to smell. The mucus plays an important part in catching the scent.[5]

Your dog can smell far better than you can. In fact, tests have shown that dogs can find a chemical in a solution diluted to one to two parts per trillion and smell something almost fifty feet below the ground. We humans depend more on our vision, and our brains have more space dedicated to seeing. Our canine companions have a larger portion—40 percent more—earmarked for smelling.

You have about five million smell receptors—compare that to the dog who can lay claim to having from 125 to 250 million, and bloodhounds have even more.[6]

Dogs' excellent sense of smell has enabled them to work to help the environment. Conservation Canines, CK9, based in Seattle, Washington, has been a part of the University of Washington's Center for Conservation Biology for more than two decades and is well-known for its training and fielding detection dog-handler teams.

By sniffing out scat with their noses, these dogs, who may have come to the end of the line as far as having a future goes, work around the world to enable their handlers to gather huge amounts of valuable information on different species in any given area, thus benefitting research on endangered species.[7]

Dogs use their fabulous sense of smell to help rescue humans and animals from life-threatening situations like natural disasters. Take, for instance, Frida, a yellow lab mix now getting on in years who has managed to save many people who were victims of terrible disasters like earthquakes or gas explosions. In other cases, she has helped to locate the bodies of those did not survive. She has worked in Ecuador after an earthquake, in Guatemala after devastating mudslides, and in Mexico after the 7.1 magnitude earthquake. Frida is part of a team of canines working with the navy in Mexico.

How can you be at peace and not have any desire? Is it even possible? For the love of your rescue dog, you can at least try. And trying is in itself a step on the path to Enlightenment. The next chapter is about how our desires cause us suffering.

CHAPTER 3

THE THIRD NOBLE TRUTH: LOSE THE DESIRE TO LOSE THE SUFFERING

How do we reconcile the quandary of not having desires with what we want for our adopted dogs? Can we allow ourselves to accept that our adopted dog is just how he or she is?

Charlotte, like the mischievous miniature poodle-pharaoh hound she is, can back out of collars and some harnesses. She can pull like a dray horse and because her neck is so thin and long, she can reach anything on the ground when she is being walked. This presented a real problem for the first several years because she would try to eat anything she could reach—and she could reach pretty much everything—and this would often result in her starting to choke or throwing up later. The "leave it" and "drop it" commands did not always work as they should. Besides, it was hard to tell if she was sniffing for a place to "go" or sniffing for something forbidden to eat, and before you knew it, some unidentifiable thing was in her mouth, which she would either drop or swallow. And the more things outdoors she would eat, the greater her chances for having serious intestinal issues and having to go the ER.

And that is exactly what eventually happened. One night, I heard what causes alarm in a dog parent's heart: the unmistakable heaving and throbbing sounds of your dog retching. Charlotte started throwing up. A lot. This was the middle of the night, so I called the emergency veterinary hospital and described what was happening; the vet tech explained that the appearance of what looked coffee grains in the vomit was likely blood and that we should bring her in right away.

It turned out my dog was so dehydrated that she had to be given a "water pack" injection under her skin with a rehydrating mix, basically an electrolyte solution. This is known as a Lactated Ringer's injection and in Charlotte's case it was given subcutaneously, under the skin. For the next week, she needed to have a special meal prepared to help her recover. This happened on at least three separate occasions and the events were always very traumatic for Charlotte and all of her humans-in-waiting. Each time she needed the electrolyte fluids.

After the third ER visit for Charlotte's severe vomiting, the vet in the ER recommended that Charlotte see the gastroenterologist to have a complete blood workup and an internal examination, either an endoscopy or an ultrasound to see what was going on with her digestive system. We made an appointment with the specialist to discuss the best options for Charlotte.

While waiting to see the internist-specialist to discuss what might be going on with Charlotte and to learn more about the endoscopy and the ultrasound, and what that might involve and show, I started to worry if it was possible we might lose Charlotte. This uneasy thought was compounded by the fact that we were at the same veterinary hospital where we had said our final goodbyes to our previous rescue dog, Alice, who had succumbed to cancer.

After talking with the specialist, I opted for the ultrasound instead of the more invasive endoscopy since an endoscopy would require her to be put under anesthesia. I reasoned we should see what the ultrasound comes up with, at least first. From the bloodwork, the vet had ruled out pancreatitis. Charlotte is part miniature poodle and they can be pre-disposed to having sensitive stomachs. Charlotte had the ultrasound and was diagnosed with having an ileus. This meant that her intestines have motility problems when it comes to moving the food she eats along the natural route of her intestines. In Charlotte's case, it seemed that the worse manifestations, such as the vomiting, could be temporary, but she needed a special diet to ensure she wouldn't have the serious complications like she had those several times before.

One of the most useful pieces of advice about judging the health of your pet was given to me by the medical director of the veterinary hospital. The two most important things to watch for are attitude and appetite. If your dog seems like her or his usual self and is acting like he or she normally does, and if their appetite is the same, then they just might be more on the side of being okay.

Being prepared for vet visits is essential to ensure the health and safety of your pet. I have found it helpful to have a running history to give whomever Charlotte needs to see for whatever reason. It saves a lot of time and having a narrative on hand that explains what has happened and when and what has been done about it can be a part of your pet's health record and is a document that you can refer to and amend over time.

Charlotte has issues with eating, chewing, and swallowing her food because of the problems with her neck, esophagus, and throat, along with the slow digestive process her body has to deal with. Even just touching her neck if she is simply standing or lying down can cause her to choke. Charlotte will always need to be given three small meals a day, spread throughout the day, evening, and night, and, until we later learned of another health issue, it was always the easily digested grain-free canned lamb, mixed with water. She cannot tolerate dry food given to her as a meal since she just tries to quickly swallow it and doesn't take the time to chew it. Balancing what to give her and when so her stomach is not too empty for too long is a little tricky.

All of this for Charlotte is what is normal for her, and it has come to be normal for all who love her. To wish it could be any other way would be pointless.

Finding the Rescue Who Will Rescue You

Say you have not yet adopted a rescue dog, are still considering adopting a dog, but you are not sure, or you know someone who has expressed interest in adopting but they are uncertain how to go about it. What is a good way to find a dog, puppy or adult at a shelter or rescue organization? I asked Dana Ebbecke[8] , behavior counselor at the ASPCA Adoption Center, about this. The ASPCA encourages everyone thinking about pet adoption to consider their lifestyle needs and living arrangements when looking to add a pet to their household. But they also advise approaching the adoption process with an open mind. Your personality and lifestyle, along with challenges such as housing restrictions and amount of time spent at home, should be explored to determine what pet is right for your household.

At the ASPCA Adoption Center, pets are assessed by certified professionals, then introduced to potential adopters based on the likelihood of compatibility. They take into account the lifestyle of the potential adopter and what they're looking for in a dog. Then they see which of their dogs would fit. "It's a bit like matchmaking," said Ebbecke, "but for dogs!" It is important that potential adopters should consider what they're looking for in terms of energy and level of commitment. Some dogs require more exercise than others, and some require more training. "Each potential adopter should consider what they're willing to invest and their capability of following through with that for the next ten to fifteen years of the dog's life," Ebbecke stated.

Sherry Woodard from Best Friends believes that bringing along a trainer or animal behavior consultant when visiting potential adoptees can help a lot[9] . Best Friends has people all across the country who are trained in reading body language and making the matches. Of course, many people are not able to do this. You just want to be sure that when you go to a shelter looking for your next pet that you are not attracted to the wrong things, such as a dog that is scared or shy. Often, as Woodard said, those dogs are going to be a project and they may never actually be comfortable doing what the person hopes that they will be able to do.

When you or a friend is looking for a puppy or adult dog to adopt, asking a trainer or animal behaviorist for advice can be helpful for finding the right match for both the person and the dog. This is a very important part of helping both human and canine get set up for success. "I think often, for me and for many other trainers and behavior consultants," Woodard said, "it's very sad when we meet the people after they've found the wrong dog. They expect us to change the dog and make it into a different dog." But that is something that they cannot do.

Woodard adds that a shy, scared, fearful dog is likely to not be comfortable in many settings long-term. "We just can't expect them to completely turn into a different animal."

Not going in with pre-conceived ideas and wishes can be a way to make this work.

Lose the Desire to Be Able to Lose the Suffering

The Third Noble Truth is the truth of the ending of suffering, *Nirodha*. Stopping the cravings will stop the suffering. If having no desires means we will have no suffering, why then is it so hard? Letting go of desires doesn't mean you shouldn't love others or shouldn't have emotions. The idea is to know that you can be at ease with what you already have and that you really don't need the things that will never satisfy you.

Letting go. Simple concept—not so simple to act on. Is it really possible to *not* have desires for what we see around us—desires that go beyond the basic needs that we all have and must satisfy to survive—eating, drinking, sleeping, having adequate shelter, and so on? These are very different from the thirst for the newest "thing." If you don't really *want* that new, newer, newest possession, you are likely to be content with what you already have.

A rescue dog doesn't think about the steak she isn't having when there is a dish of her usual food in front of her. She is hungry and she eats her meal and then (hopefully) is full. She has no unsatisfiable craving for what she has not been given.

But how about you? How can you be at peace and not have any desire? Is it even possible? For the love of your rescue dog, you can at least try. And trying is in itself a step on the path to Enlightenment. When you are with your dog, *be with your dog*. When you are walking your dog on a city sidewalk or in a park, try to not think about what you might be doing later in the evening, or where you should have dinner next week. Wanting more or wanting something other than what you have is a trap and you will never be satisfied. That is what makes desire painful so losing the desire will diminish the inevitable pain.

> *"Many of us crucify ourselves between two thieves—*
> *regret for the past and fear of the future."*
> *—Fulton Oursler*

What is Enlightenment, really? Is the journey in itself enough of a reward? Is at least trying to stay on the path to Enlightenment in some sense a kind of Enlightenment? Is it enough for most of us to let the journey be the reward?

Another problem we humans have is letting our thinking get in the way of our trek to becoming more aware of and content with what we have. Thinking too much and overthinking causes problems with trying to let go of your desires.

And if we think too much, we run the risk of becoming stuck and not doing anything. We think so much that we cannot make a choice. Again, sometimes regretting things that happened in the past as well as fear of what might happen in the future can lead to being frozen in place, and not making a choice or not making a decision. But, as we keep hearing, not making a choice is itself a choice. It does not matter if we are considering big picture changes or smaller changes or whether we are ruminating over small choices. Not making a decision may seem easy at first but in the long run, the results of that non-decision will happen. Life happens. Anything can happen. Everything can happen.

Life is full of little unexpected events and as human beings it can be extremely hard to simply accept what happens. Peace lies in acceptance; not resisting what happens and the struggle to get to this point is worth the effort.

How Have You Handled Suffering?

Not accepting that there is suffering results in the greatest suffering. Fighting against the acceptance of suffering only reinforces its influence. So, accept that there is suffering, and you will diminish its power. And since perhaps the most emotionally debilitating suffering is the suffering caused by desire, letting go of the "I want" is a start.

Now might be a good time to pick up your journal and find a quiet place to sit with or near your dog and ask yourself these questions:

- Are you happy with yourself?
- Do you think a lot about being dissatisfied? Wanting to improve is a positive thing, so what is wrong with trying to improve? Being satisfied and content to live with what you have now is a more positive (and more productive) way to try to improve. If you start out by being dissatisfied, then you will always have that feeling of dissatisfaction.

- Do you find yourself wanting more and more things? It is so easy today to see the newest fashion or gadget or car available with a simple click. It is called the internet of things for a reason.
- Are you jealous of someone else's acquisitions or success?
- Are you frustrated because you don't know what you really want?
- Are you disgruntled about the fact you are disgruntled?

While we all experience these at some point and to some degree, how can we let it go?

Write down your answers to some or all of the above questions. Then, ask yourself how would your dog answer these questions if he were able to talk? Think about it. Write down not only how your dog might answer the above questions as they apply to him, write down what your dog might have to say about how you answered the questions. The next chapter will be the threshold to finding ways to lose unquenchable desires.

CHAPTER 4

THE FOURTH NOBLE TRUTH: FOLLOW THE EIGHTFOLD PATH TO LOSE YOUR DESIRES

How do we lose our desires? How can we learn to let go? How can we live more comfortably with some issues we may be having that we find hard to resolve? Our dogs can teach us. But we have to pay attention.

Dogs in general, and rescue dogs in particular, have been used for years to help people who are struggling with physical and mental health issues. What is it about rescue dogs that makes them so able to do this?

People who are suffering from a mental illness can find comfort from and empathy with a rescued dog. One person explained her experience with this. Rita is living with a chronic, major mental illness, and she has benefited from having a companion dog, and especially a rescued companion dog. As she explains, "I feel like rescue dogs in particular connect with humans who have special needs and have been through a lot in life because the dogs themselves have most likely been through a lot." Rescue dogs just seem to understand what their human is going through, she believes. "I can say from experience that I feel like my dog won't judge me if I'm having a particularly hard day," Rita says. "She is just there for me when I need her and ready to be hugged, and in turn, comfort me. I also think that dogs who live with humans with special needs develop a sense of their person's emotions and therefore instinctively learn when to offer comfort."

Dogs don't judge a human based on their condition or the place they might be in; they just reciprocate what they receive, which is—hopefully—love. Even in cases where the dog has not been surrounded by love but has been abused, the dog still looks to his or her human. In Rita's case, her dog doesn't see her for her mental illness; her dog just loves her because she shows her dog love.

Rita thinks that this is because rescue dogs also tend to rescue their people. All dogs are great, but rescue dogs in particular tend to bond deeply with their humans because they most likely had to overcome a great deal to get where they, hopefully, ended up—in a safe, loving home. "I find that dogs tend to gravitate toward people with a special need because they somehow sense the human needs 'extra' comforting," Rita adds. "Dogs are amazingly sensitive creatures who show humans so much love and devotion and don't ask for much in return."

She isn't the only one who realizes this. Pets help people with mental health issues establish routines and get exercise. Pets can offer a welcome distraction and bring a sense of control to their lives. The journal *BMC Psychiatry* published a study that concluded that pets should be deemed a key component and not a negligible source of support when it comes to managing long-term mental health issues. Pets should be included in the arranging stages and the implementation of mental health care. Clearly pets need to be more than an afterthought when it comes to helping people with a mental illness live a better life.[10]

Another view of how rescue dogs have come to be used to help humans in need, whether a veteran with PTSD, a person in prison, or people in hospitals or nursing homes comes from the ASPCA. Why is it that dogs in general, and dogs who have been rescued specifically, are so good in this role? According to Dana Ebbecke, behavior counselor at the ASPCA Adoption Center, studies show that dogs provide more than just companionship to the humans who care for them. Dogs also help people live longer and healthier lives and contribute to their caretaker's physical and psychological well-being. "Dogs are social animals, as are people, which helps to facilitate bonding and the flow of happy chemicals in the brain," said Dana. "Adopted animals are especially helpful for people in need because they provide a great metaphor for trust and overcoming adversity."

The ASPCA is a Community Partner of Pet Partners. These organizations are allied because they both promote the fact that every animal is an individual. "Appearance or pedigree is no guarantee of temperament or aptitude," said Dana. "A significant number of currently registered therapy animals in New York City were adopted from rescues and shelters. We also have animals with disabilities and even cruelty survivors that go on to serve their communities with their humans." When an adopted pet becomes a registered therapy animal, according to Dana, it helps dispel two common myths:

Homeless animals are damaged goods. No, they were often loved and well cared for before their humans had a crisis. Most shelter animals are not victims of cruelty or neglect, just unlucky.

One needs to raise a dog from puppyhood in order to "mold" them into their future job. This is not true. Training from an early age helps, but these special animals are born, not made.

Taking care of a dog can be emotionally rewarding, especially for someone who has very little control over his or her life, because that dog is dependent on the person. And the human has unconditional love. This sense of responsibility can be very healthy for people who have undergone psychological trauma or who have had certain freedoms taken away for whatever reason. And dogs can teach children compassion and empathy, taking care of somebody who's even smaller than they are and yet needs them desperately.

Marion[11] , (not her real name) a business professional, gives the example of her son who was recovering from depression and had to leave college. He took a job walking dogs, and she believes that is significant because they had a dog when he was growing up. They have always had dogs, but her son's first relationship was with their dog rather than with people.

Her son walked the dogs and took care of them at home, and then he worked at a shelter for rescue animals. He was able to transition back to working with computers, and she credits his work with dogs and focusing on their care as the impetus for that.

Following the Eightfold Path Is a Way to End Desires

Caring for dogs was a way back for Marion's son, and caring for a rescue dog can help any of us to keep our feet on the path to a more meaningful life. Following the eight steps, *Magga,* can lead to losing our desires, and it is the desire that causes much of the frustrating suffering we feel.

One way to silence the busyness that plagues us is to set aside time each day to meditate. The purpose of meditation is to put an end to the zig-zagging and aimless wandering of the mind or even distracting deliberate thoughts. It can help to quiet the mind.

Meditation helps and has proven to benefit humans on many levels, from the physical to the metaphysical. Meditation can take many forms. Just sitting quietly for about fifteen minutes a day can have a profound effect on how you perceive the world—both the outer world and your inner world. There have been quite a few studies that show meditating can physically change the brain. Meditating on a regular basis can help aging brains to stay more pliant. Meditation can shift the focus from what is known as the Default Mode Network (DMN), the area of the brain that is the culprit for you being distracted and too self-focused. Meditation can help mitigate depression and anxiety.

Meditation

In Buddhist tradition, meditation makes up the second element of the threefold path. There are quite a few interpretations of the Buddhist path to spiritual awakening, but the threefold path is largely understood to be the most fundamental. There is no one path. Each way is a part of another path. The first training in the threefold path and the necessary foundation for spiritual growth is ethics (*shila*).

The second training is meditation (*samadhi*). The idea with the first training is to prepare yourself to be in the right frame to be able to do the second training, meditation. If you are conducting yourself in an ethical manner, then your life is uncomplicated and your conscience is clear and this is the best preparation for meditation.

"Walking meditation is really to enjoy the walking—walking not in order to arrive, just for walking, to be in the present moment, and to enjoy each step."
—Thich Nhat Hanh

Focusing on walking your rescue dog paves the way to the next steps, the Eightfold Path. For example, walking can be a form of meditation and so can walking your dog. Find a quiet path and as you are walking your canine companion, try to not think about anything else but your dog and how she is walking. Of course, you have to be aware peripherally of your surroundings for safety, but concentrate on feeling the rhythm of your steps and your dog's stride. You both will fall into a rhythm and, barring an unexpected chipmunk, rabbit, or squirrel crossing your path, you both should be able to fall into sync. This is a way to empty your mind of anything that has nothing to do with walking your dog at this moment.

This is meditating on the move. Your mind is still while your body is moving. You and your dog are going places, getting fresh air and exercise, and at the same time, you are clearing distractions from your mind and focusing on not focusing.

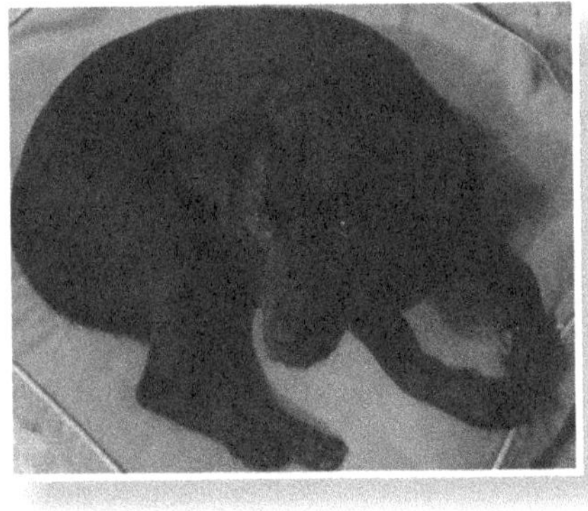

PART TWO: THE EIGHTFOLD PATH

Mind your thoughts, as they become your words.
Mind your words, as they become your actions.
Mind your actions, as they become you.
—Buddha

CHAPTER 5

THE FIRST STEP: THE RIGHT VIEW

It is not easy to see things as they really are—to not let ourselves be conned or seduced by a near-perfect object, experience or feeling—whether it is the perfect place to live or finally getting the car we always wanted, or something as intangible as the love of a person. The happiness seems so permanent and deep. How could it ever not be there? So, to make sure we never lose the brilliance of what we are feeling, we hold tightly to the joy we think we have found—so tightly that we may not even notice when the joy begins to dissolve and seep through our fingers.

Charlotte eventually returned to her lively, smart, and inquisitive self, and we had to find a new vet and this time we wanted one a little closer to home. We still did not want to go back on a regular basis to the hospital where Alice had spent her last hours. That wound had not healed, and indeed, would take a long time to form a scar, even with the irrepressible enthusiasm and boundless energy of Charlotte.

She was our happiness, and she was our present. But letting go of the past that Alice represented was proving to be very hard.

Nothing is forever. There is some truth to the observation that getting a puppy is a countdown to sadness. Every beginning has an end. Enjoy the beginning and all the time with your adopted dog. Accepting that nothing is permanent, no matter how tightly we cling to it, is a step in the right direction. Your dog will make sure you are aware of how things change constantly and lead you to accepting these changes—some tiny, some pretty big.

Sometimes, as with what happened to Alice, a fatal, unforeseen illness can come out of nowhere and knock the breath from you and really set you off course on the road to Enlightenment.

The sadness and the grief that wash over you when your beloved pet dies reach into every part of your daily life. Like any grief over the death of a loved one—human, canine, feline, avian, and so forth—closure never really comes. We carry with us always the heavy hollowness left by the loss of the loved one.

Don't grieve. Anything you lose comes
'round in another form.
—Rumi

When Alice was diagnosed with a terminal illness, we tried to do everything possible to prove the diagnosis wrong. I had adopted Alice as a young puppy at a time when we welcomed the source of comfort and focus she provided.

When we first spotted Alice on the rescue dog site, we knew she was special and smart. Her eyes had a depth that showed her old soul within. Every dog, pup or senior, deserves to have a loving home, but something about Alice just pulled us in. You could tell there was something behind her beautiful and sad eyes. When we met her at the foster home, she looked up, turned around, and walked back to the little pool that had been set up for her and another puppy, as if she was resigned to being overlooked.

Alice had never been in love with food, even as a very young puppy. She could easily take it or leave it, no matter what form it took. We kept a supply of tubes of nutritional paste to give her in between her mostly untouched meals. And yet she always ate just enough to not trigger any alarms and when she went to the vet, she always came back with a clean bill of health. She had been a parvovirus survivor, but she seemed fine.

Alice loved going to the vet. Some dogs have to be coaxed or carried (like Charlotte has to be) to go with the vet or the tech, but Alice would always trot off happily. Looking back, I wonder if there was a reason she loved going to the vet. Did she somehow know something about her body that no one could discover until it was too late?

I brought her to the vet twice that week, the week when her not eating seemed to have become much worse. The second time, I told him about how she stopped eating her food altogether but would eat some of her favorite treats. I had brought some with me to show the vet, but she refused to eat even those at that time. Alice really liked this vet a lot. They seemed to have a good connection. I asked him what could be wrong? He said it might just be her mood or….

"Or what?" I asked, tentatively. His silence was a little off putting. Then I started to get really frightened. He said he would find out. He took her to a bigger exam room at the hospital. I don't know how long I was waiting for him to come back to tell me what was ailing Alice.

Another doctor I did not recognize came back with him. The other doctor turned out to be an emergency internist and surgeon. She was dressed impeccably and that, along with the determined clack of her heels on the hard floor, irritated me almost irrationally. Her manner was a bit brusque, and the delivery of her words left no room for any kind of more optimistic interpretation. Alice, they discovered, had metastasized lung cancer and there was nothing medically that could be done.

You know how the room spins, and you feel you have been punched so hard the air gushes out from inside you and you are left feeling completely numb and cannot even find the energy or the will to stay standing? How could this have happened? There had to be some other explanation. There had to be some error with the x-ray itself or the reading of the x-ray. It could not be possible. She came regularly for checkups. Where was this coming from?

The reaction to devastating news like this is similar to the reaction to the actual death of a loved one. First, there is denial. Nope. This had to have been a mistake. I had to call other family members to tell them the terrible news. Alice wasn't just sick with some simple canine ailment. Alice was dying from metastasized cancer.

Then there is anger. Lots of anger. Anger for not having discovered this sooner—a lot sooner—when something could possibly have been done to prevent the cancer from spreading from where it had started. What had been the point about being diligent about taking her to the vet for regular checkups? Any time we thought there might be a problem, like she was favoring a leg or starting to limp a bit, we brought her in to be examined. Constantly searching for answers, I asked myself if I was somehow remiss. Did I not do enough? Should I have pushed harder for more thorough and complete exams and medical tests? What had I missed? Was it my fault? Could I have prevented this dear, sweet, brave, and strong girl from getting this death sentence at the youngish age of six-and-a-half years?

You will not be punished for your anger.
You will be punished by your anger

From that day, the journey to Alice's final destiny was one fraught with anger, denial, intense research into what else could be the cause of her deemed-fatal cancer. We made sure she was comfortable and had everything possible to help her with her breathing when it became difficult. I could not let her go. How did this happen to her? Why did this happen to her? Why were we being blindsided by losing our loved, wonderful and beautiful dog? How could this be in any way fair?

No matter what emotions churned through my body and my soul, Alice needed to be comfortable during her final weeks. I will never forget it. It was Christmastime and everything that was happening around us was in stark contrast to what was happening inside our home. Alice was dying and no amount of pathetically cheery holidays songs would make that better. We kept pushing for more tests and hoping for some kind of hope but none was forthcoming. Alice was now having difficulty breathing and we helped her with as much as we could, but we could see we were losing her. The light was no longer there in her eyes and when we took her outside, she sniffed the air as if it was not really familiar to her. I gave her water but she could not keep it down.

I brought her to the vet again, and he said it was time to come to terms with letting her die in peace and to end her suffering. To say it was difficult to hear this, let alone make the decision about what to do, is a deep understatement. If the goal is to learn to accept events rather than fight against them, then I was failing miserably. I could not and would not accept that we needed to decide about letting Alice breathe her final breath. About being there with her, watching her, and holding her as the life left her body. Where does that life go? One moment she was running happily and carefree, another moment she was diagnosed with a terminal and fast-acting illness and the next, the life, all of her energy, would be leaving her body. Where would that energy go?

If it is true that energy cannot be created or destroyed, what did that mean when the energy from a living entity left that person or animal? Where would that energy go? Is the energy the spirit of the living person or animal? Did that mean that Alice's spirit would somehow be with us always, and not just in my heart? Did it mean that Alice's spirit could be reincarnated in another living creature?

Maybe the next dragonfly I would see in the spring or an unexplained presence of an unfamiliar bird would be the vessel containing Alice's spirit. But you never really come to terms with having to euthanize your beloved dog. It is an act that there is no going back from. It means her life will be gone forever. But to end her suffering, it had to be done. Being with her in a small room with the vet and allowing this to happen will forever change me. It is a memory that will stay with me always. Whatever the reason they were composed, the Rolling Stones' "Paint It Black" and James Taylor's "Fire and Rain" convey some of the feelings that circle slowly and sorrowfully in the soul after the loss of an important presence in your life.

Veterinary medicine has made many advances since Alice was diagnosed with metastasized cancer those years ago. Today, it is recommended that your pet have a wellness check that incudes wellness screening tests. There are four principal types of wellness testing that can be beneficial for a dog's overall health: a complete blood count, a biochemistry profile. urinalysis, and a thyroid hormone test.[12]

Each of these categories can be broken down to include even more pinpointed categories. If the dog is younger, as Charlotte was at the time we started getting these tests, it may be fine to have fairly basic testing. But since Charlotte also had many health issues that needed to be managed, she would need more extensive testing.

If your dog is on the older side, however, tests that are more thorough are a good idea. You might want to add chest or abdominal x-rays to check the size and appearance of the internal organs (heart, lungs, kidneys, liver) or x-rays of the skeletal system to look for any bone or joint changes that may be indicative of a degenerative condition.[13]

If only we had known about this for Alice when she was going for her checkups. But, here is the thing. We know what we know when we know it. We knew what we knew when we knew it. We cannot wring our hands and exhaust our minds with regret about the past. We can't think about how we might have done something differently had we only known then what we know now. It doesn't work that way. Hindsight gives you clear vision, but only into the past, something we can never have into the future. So, stick with what is going on right now in front of you. Enjoy every moment and experience you can share with your dog. Love him or her for what she or he is at this moment.

"Do not dwell in the past, do not dream of the future, concentrate the mind on the present moment."
—Buddha

When we take Charlotte for her wellness checkups, which we do twice a year, we have complete blood work done so that any hidden ailment or illness that might be brewing can be spotted early. It is not foolproof, of course, and even this, along with a conscientious vet and helicopter pet parenting, will never guarantee that Charlotte will never have to suffer down the road with a cancer or other illness. But we do what we can.

While Alice was here, we became a part of each other. We were responsible for her care and she gave us unconditional love in return. When Alice became irreversibly sick, it was up to us as a family to help her in her pain. No one wants to be faced with this decision. It is an impossible decision. But not making the decision is, in itself, a decision.

Have the Right View

The things that can go wrong in life never cease to amaze me. Just when you think you have a life-altering event under some kind of hapless control, another mental and emotional brick is thrown at you head on.

This is where understanding that we never really have control of things outside ourselves can help ease the frustration. Zen and the rescue dog can go hand-in-hand, but the road to peace is not a straight line and there are many detours along the way. And there really is no end to the road anyway. It is not so much what happens to us in life but how we react to these events. That, my friends, is easier said than achieved.

When I adopted Alice, one of the first things I did was to get pet health insurance for her. If it could work for humans, it might be work for animals. I have become a firm believer in having health insurance for pets. I have never regretted having it but I have regretted not having it.

Many vets recommend not only regular health checkups for your pet but also having pet health insurance. Having health insurance for your dog is at the top of the list when it comes to how to ensure a better relationship with your dog. "The biggest thing, practically speaking, is health insurance, since facing an expensive health crisis can have heart-wrenching outcomes," one vet told me.

This vet has seen many pets with varying degrees of medical problems and notes how strong the bond can be between the pet and the pet parent. When a pet dies, it can profoundly affect all of the humans who have come to love the pet. These bonds are, as she says, as powerful as any human to human relationships. "People tell me frequently that they grieve more for the loss of a pet than for a human family member."

The ASPCA website offers some counsel for those managing the care of a very sick pet and those who are suffering the loss of their pet.

What happens to another pet in a household when one pet dies? Pets, too, can feel grief. A recent study looked into this very topic and investigated how other animals in a home could be affected by the loss of another pet. The conclusion is that the behavior of the remaining animals in the house changed in various ways, for example, the living pet became more affectionate or changed his or her eating habits.[14]

Animals can often exhibit the same signs of grief as do humans. Studies like this are important because they help humans to better comprehend how to better care for their pets. There are ways we can help our pets when a person or another pet they have formed a bond with suddenly or lingeringly vanishes. Animals, like humans, may want to either be alone or crave more than the usual amount of attention. It is best to respond in a way that will give comfort to the animal and alleviate his or her needs, within reason.

Try to understand what your pet is trying to show you. Pay close attention to their behaviors. If you pet seems to be needier and follows you around, go the extra distance to spend more time with her. On the other hand, if your pet wants to be left alone, it is better not to come on too strong. Let him have a bit of distance but also try to encourage him to ease into some of the activities he always loved.[15] It might be hard to pay a lot of attention to the pet who is sad because you both have lost a beloved friend, but doing so might help both of you to heal together.

When we lose a beloved pet, it might be a natural reaction to wall ourselves off from the possibility of a hurt like this every happening again. If I adopt another puppy or adult dog, it is inevitable that I will suffer through another loss. But to not want to adopt again is not a way to honor the pet you lost.

This is a very sensitive issue and a very personal one, and Sherry Woodard had some thoughts about this. "I think this has been a challenge for me because sometimes you hear people say, 'I'll never have another animal because it [the loss] was so painful.' Some of those people are wonderful, wonderful homes that we can't lose. We need them. We can't lose them, so somehow [we need] to get across that fact that they were loving, amazing homes and we have animals waiting in line to be loved and to have a home."

Try to say yes. To adopt again means to have another chance at caring and loving another animal. Adopting again means being able to receive love from another dog. The ASPCA concurs. "Adoption can be a rewarding experience for both the human and the dog," stated Dana Ebbecke. The dog gains a loving home and a safe environment. The human gains companionship and the knowledge that they are giving the dog a loving home and saving the life of another animal. "Adopters allow the shelter to move their resources and attention to another animal and continue making an impact on the welfare of animals in their community."[16]

Sherry added, "We know it's painful, and we know it's never supposed to be easy to lose an animal or to go through the pain. The pain is going to be there whether it's an animal that gets sick young or an animal that we have for however many years."

Sherry has a dog who is fifteen years old and she has had him since he was three days old. She knows what the future may bring will not be easy. However, she said, he's not going to be the last. "I wouldn't want to live without someone in my life that is a non-human family member." She added that what they bring into our lives is so precious and so valuable; we continue to learn from every single one of them every day. “We can’t give that up,” she said, “and we also can’t be selfish, especially at a time when there are so many of them waiting for us that we have to give more and open ourselves up more.”

No one saves us but ourselves. No one can and no one may. We ourselves must walk the path.

When Joan Rivers’ Pekingese Max died, one of her friends threw a funeral and it was an occasion suffused with humor. Rivers told the *Chicago Tribune* that the comical sendoff helped her to manage her grief at losing her beloved Max. Humor and being super busy were her ways of coping.

Rivers recommended adopting another dog after losing a loved pet. As she said in the article, she loved Max, and he knew she loved him. “But he’s gone now. I try to fill the void as quickly as I can; it’s good for me. But it’s also good for whatever dog is brought into my life, since I only get rescues.”[17]

Help Finding the Right View

The sadness of losing a pet never really goes away, no matter the circumstances around the loss. Whether it is a long time coming or sudden. It is still loss. It is hard at the time to see how anything like that can be even remotely accepted.

I look over at my now healthy, energetic, and still very puppy-like Charlotte and know that this, too, will happen at some point with her. Hopefully not until she has enjoyed a long life, but nonetheless, the specter is always there. The Sword of Damocles hangs above. It may be obscured from view most of the time, but it is there. It is sometime feels a bit like having cherophobia, or the "sadness of a happy time," although that, of course, would be self-defeating not only for me but for Charlotte. Not enjoying her fully while she is here and helping her live life to the fullest would be shortchanging both of us.

No matter how close we hold happiness, we can't keep it from changing and evolving into something we might not want. If we can recognize that change is going to catch up with us even if we try to keep one step ahead, then we can be more accepting of the things we don't want in our life because they are too painful.

The only cure for grief is action
—G.H. Lewes

One way to work through grief is through work. Keep moving forward and keep working. And working on a journal during a time of grief can help. When grief and sadness weigh on your soul, it can make you feel alone and disconnected. Other people can offer sincere condolences and even commiserate to a point, but after a certain amount of time, there is only you and the heaviness of your sorrow.

Writing in a journal as often as you can and putting anything down on paper and detailing any and every memory of the person or pet you have lost can help. Use a freeform style, don't be concerned with editing or second-guessing what you meant to say. Just get your thoughts into your journal. It also is a way to have a permanent record of your thoughts that you can turn to whenever you need to remember.

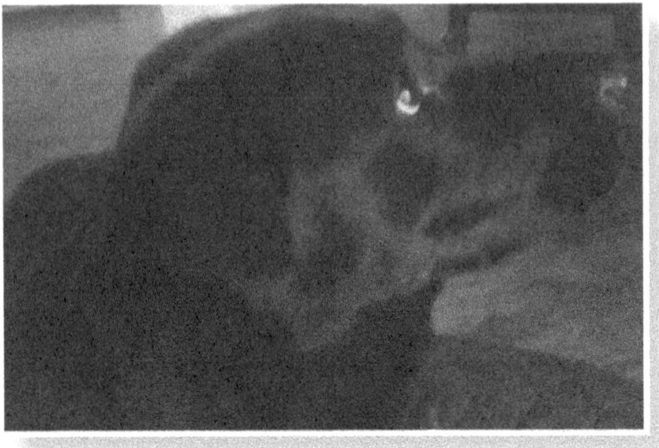

CHAPTER 6

THE SECOND STEP: THE RIGHT INTENTION

Are we always honest with ourselves about the motives behind our actions? Is it even possible to always be straightforward about why we are doing something or saying something? As humans, our mind can be very clever about convincing us that we are really saying what we mean and we are really doing what we said we would or did do. When, honestly, the truth is somewhat different.

We want our dogs to be healthy and phobia-free, but some dogs have ingrained fears that we just have to work with. Pet parents need to know when and how to help their dog, but we also need to learn that there are things we just cannot change and we must accept them.

When Charlotte first came to live with us, it was winter and there was quite a bit of snow on the ground. Taking her outside in a foot of snow so she could "go" required patience and warm clothing for both human and canine. She spent much time sniffing for just the right place and then she dug a hole in the snow to accommodate her needs. Interestingly, when warmer weather came and there was no more snow, she would find it difficult to get just the right spot to go. She had become so used to the snow and digging the holes that going just took a lot longer.

Several months later, Charlotte was making progress becoming more comfortable and confident, but she still had definite anxieties when it came to going out and about. Since Charlotte has a very long neck and strong legs, she can effortlessly and quickly pull hard from one direction to another when she is being walked, and this was a real problem for the first several years. It took some time before we found just the right combination of collar and vest to keep her secure when she went out for walks.

I enrolled her in a beginner's training class, hoping to give her some confidence as well as helping her to benefit from some basic training from a professional dog trainer. But first she had to actually get from the car to the building and then on into the classroom. It was clear that this was not going to be easy. Carrying her from the car was what worked.

Since walking with Charlotte down any hallway presented special challenges, getting her into the building and then from the entrance down the hallway to where the class was held proved to be a special struggle. She would make herself flat and one with the floor, like she was doing a combat crawl, only she wasn't moving. Someone had to kneel at her side to coax her along.

Any nervousness the pet parent might have could translate to the dog, since they are very attuned to our emotions, even feelings we may not realize we have. There have been studies about this concluding that dogs are intuitive and that they can easily tell we are stressed out just by reading our body language and even by using their remarkable sense of smell. If someone is anxious or nervous, they tend to sweat, and the dog can smell the apprehension.

Some time ago, there was a television program that purported to show how having a dog might not really deter someone from breaking in. Since the whole scenario was staged, as it had to be, the dog knew that the "intruder" wasn't a threat. The dog likely could tell that nothing was really wrong by picking up on the actor's scent and demeanor. A real intruder would likely be giving off all sorts of signals for a dog to catch and react to.

With Charlotte and the training class, I did my best to remain calm and to get my physical self in line with my mental self and vice versa. If I could remain cool and confident, then Charlotte just might be more confident as well. If I could control my own emotions and anxiety, Charlotte might feel better about what was a new experience for her.

Now we were at the point where I was getting a little concerned that she might not even make it to this first class, which I knew she would most likely really enjoy once she got inside the classroom and met some of the other dogs. Charlotte was a super intelligent dog and she was a fast learner. She just needed to be comfortable in her own fur, though, to be the self-assured dog she could become.

How would we be able to accomplish these goals? We met the trainer in the hallway and even with her coaching, Charlotte would not walk through the hallway on her own. She refused to move. She took the paw-plant/donkey-stance maneuver further and used her skills as an accomplished floor hugger and splayed herself out on the smooth floor so she could not be moved.

But we picked her up and made it to the class, and Charlotte proved to be a quick learner and got along great with the other puppies. Every so often, a fear would manifest as when during one lesson, the trainer and the assistant gave each of the "students" large metal water bowls. Charlotte panicked at the sight and the sound of the bowls being placed on the floor down the row of "students" and their pet parents. She would not go near it.

But overall, Charlotte learned many commands during the class and with consistent practice at home, what she learned was reinforced so that it was all second nature to her. She still loves to make a game out of learning commands and she was always a very eager and quick learner. But as with all dogs, practicing regularly is key to the dog being more consistent with the learned behavior. One way to reinforce the training is to make a game of it.

One trainer recommended an activity called the "Come Here Game." All it takes are two people and the dog, and some very small pieces of the dog's favorite treat. One person stands on one side of the room with some of the treat pieces and the other person stands at the other end of the room with the treats. The dog ideally is in the middle and one person says the command "Come" or "Here," whichever word the dog understands to mean to come to the person saying this. Then when the dog obeys, he or she gets a piece of the treat. The instructions get more complicated as the game continues. Once the dog comes to the person giving the "Come" or "Here" command, that person then can add the command of "Sit," or "Down," or "Shake hands," or another command and the dog gets the treat when he or she successfully complies. A more complicated command might be something like having the dog "Stay" in place while the person leaves the room and comes back. Another exercise that gives the dog a bit of a workout is the "puppy push up." The person gives the commands "sit," "down," "stand," or "up," or whatever word the human uses for "stand up," repeated in fast succession for several rounds. The dog has burned some energy and will really enjoy the well-deserved treat after.

If your dog gets so good at these exercises that she automatically does them as if by rote, anticipating your commands, you can add some new commands or switch them around to keep both of you on your toes. In any case, making a game of learning makes it more enjoyable for both person and canine.

One of the games Charlotte loves is hide-and-seek. If I make a motion that signals I am going to leave a room, she barks and tries to race to beat me before I can hide, even if I am just trying to go from one room to another down the hallway. She gets very excited and runs to catch me before I even leave the room. Hide-and-seek raises the confidence of the seeker because the game shows the participants that they can disappear from the sight of each other but they will then rejoin each other. It is soothing in a way since the seeker will see the disappearances as brief. This game also gets her to start to go down the hallway that she used to fear so much. It is a fun way to teach her to let go of her fears.

Playing games with your dog is a way to reinforce the connection between you both. And rescue dogs have proven they are natural at forging a strong bond with people who are struggling in some way. The Research center for Human-Animal Interaction (ReCHAI), a collaboration between the University of Missouri Sinclair School of Nursing and College of Veterinary Medicine, is focused on the benefits of human and dog interaction. Rebecca Johnson is the director of ReCHAI and we talked about how rescued dogs are a good match for people with special needs.[18]

Most of the dogs in their research programs are from animal shelters. First the dogs go to the prison to be trained to the canine good citizenship level by the offenders, then the dogs that do best are selected to be K-9 service dogs for military veterans. The next step is training the dog to make sure they are equipped to help their veteran in ways that that veteran needs the assistance.

ReCHAI has placed shelter dogs with veterans with fairly complex needs for physical assistance type skills and emotional assistance for disorders such as PTSD. The dogs can also help in terms of providing personal space and alerting when someone's coming up from behind. These are skills that dogs can do very well, with proper training of course. The program has been very successful.

History of the Study of Animal/Human Interaction

Human/canine interaction has been going on for centuries, and it was in the 1970s that research started on human animal interaction (HAI). Then in the 1980s, the National Institutes of Health (NIH) held a technology assessment panel with leaders in the field at that time to discuss the state of the research, how good it was, and what could be drawn from it and then what could be done to facilitate it in the future. The conclusion was they were off to a good start, but that more research on human and animal interaction on a national level was needed.

As of 2009, the Mars Corporation formed a public private partnership with NIH through the National Institute of Child Health & Human Development (NICHD) and more funding then started going into the research. The only way to have good science about how canines and humans interact is to have good quality research and that takes funding.

Shelter dogs have proven a benefit to the community in more general ways as well. For example, through a study ReCHAI did about dog walking in the community, they were able to demonstrate that anybody who dog walks is going to be able to get more physical activity in their life in general just by doing the dog walking for the animal shelters. While anyone walking their own dog gets exercise, people walking shelter animals on Saturday mornings get more exercise, even just walking the dogs for one mile. Shelter dogs benefit as well, of course, by having someone helping them to get exercise and get their energy out.

Rebecca emphasized that the staff at ReCHAI want people to not overlook the fact that shelter dogs make wonderful pets. Additionally, as ReCHAI has found, shelter dogs also can be a good fit with people who need a service animal.

Caring for a shelter dog can help you dissipate your own desire and longing. And as we have seen above, rescue dogs can help humans with the process of unloading the burden of wanting more, the hunger that can never be satisfied.

Have the Right Intention

I know from living with Charlotte that dogs can have phobias. It seems that some of Charlotte's behaviors at times are beyond phobias. Can dogs have autism? Sherry Woodard said, "We don't have proof that they can't be autistic. We do have proof that they can have PTSD. They can suffer just as people can."

It is hard for people to change their behavior and so it is to be expected that it is probably even harder for an animal to change his or her behavior. It is harder for the animal to *understand* changing behavior, so expecting them to be able to let go of things is harder.

How to Work with your Dog and Her Phobias

If a dog has phobias like Charlotte—for instance, a fear of walking through narrow spaces like hallways—Sherry recommended setting up a scenario that would help the dog get used to walking through a hallway or narrow spaces without having the dog actually walk through a hallway.

Sherry suggested something like placing cushions or something that's really soft and something very safe in different settings and arrange them to form a small "hallway." Coaxing the dog to go down the hallway-like space created by the cushions can help her to generalize that she can do this in a lot in different places. Using treats as an incentive to reach her goal can help with this exercise and then this can translate into her being more comfortable walking through more different types of spaces in general. The phobia could be a lifelong challenge, and it may be that the dog never can let go of the fear, so the human just has to be understanding and willing to adapt.

Sherry advised trying to help the dog who has a fear of walking through narrow spaces also by thinking about other elements around the issue. There could a lot of different things that are playing into the dog's behavior in small spaces. Maybe a different type of lighting or a different kind of rug or carpet or having no floor covering would help. There could even be an acoustic element that is involved. The root of the problem can go back to something that happened to the dog while he or she was in a narrow space, such as something falling on her or him while they were in that space and could not avoid the falling object.

For a dog like Charlotte, who is inexplicably afraid to walk face-forward into some rooms but is fine walking regularly into other rooms, Sherry advised also working with treats so that the dog can be distracted and at the same time she could have really positive associations to being able to walk through the narrow space or walking into a specific room. The treats would act as a positive reward for going down the hallway or walking into a certain room and would distract the dog enough so that she would make progress at least part of the way. Also offering a special toy or a new toy she doesn't expect would be distracting enough without adding extra calories.

If the dog is afraid of boxes, you might try putting a box someplace in an area the dog feels safe in and let the dog slowly explore that a bit at a time. It may prove too traumatic at first but taking the box away and trying later or another day might work. You just don't want to make your dog's comfort zone uncomfortable.

Dogs with Special Physical Needs

Perhaps adopting a dog that has been rescued and has special physical and/or emotional needs can bring the human even closer to enlightenment. There are many dogs and other pets who are blind, deaf, partially paralyzed or more who need homes with people who have the time and resources to go the extra distance.

Organizations such as the Blind Dog Rescue Alliance in Seymour, Connecticut; Pets with Disabilities in Prince Frederick, Maryland; Special Needs Animal Rescue & Rehabilitation in White Plains, New York; and Pets for Patriots in Long Beach, New York; Lakeland Animal Shelter in Elkhorn, Wisconsin; and Heath's Haven in Post Falls, Idaho, are just a few that have animals with special needs waiting to be adopted. These and other organizations work to help as many physically challenged pets as possible find homes and to help educate everyone about the special joy that can come from adopting and caring for a dog with special needs. Dogs with special needs—physical and emotional—can give back so much to the human caring for them.

No matter how much you love your dog, caring for a pet who is always ill can be very draining for the pet parent. The fast depletion of energy and financial resources can be very stressful, and the pet parent can eventually become debilitated themselves, not unlike caregivers who look after people. Mental and physical exhaustion are always close at hand. It is important for the pet parent to have a network of friends or family who understands that their dog needs this extra care and that they need the support of others so that both the human and their debilitated pet can have a less taxing life. As with caregivers of humans, pet parents of very sick dogs can suffer from caregiver fatigue. They can experience depression and forget to take care of themselves while caring for their loved pet.

The Right Intention

Finding the Truth Behind Your Actions

What is behind your actions? Are you being selfish and controlling? Does what you are doing benefit your adopted dog, or are you just doing it because it is convenient for you? It is better to take the long view about what is really best for your dog. Your adopted dog may come with some habits and that may be hard for you to accept at first.

How You and Your Eccentric Dog Can Bond with Each Other and Be More Centered

You likely have heard of Doga, doing Yoga with your dog. Dog Yoga has been around for some time and this might be something to try with your dog. While there are limitations on what your dog can do in the way of poses or positions, he or she can figure by watching you that something interesting might be going on. Your dog might be a natural when it comes to the Downward Facing Dog and Upward Facing Dog yoga poses, but you can try some other stretching exercises with her to see if you can get her interested in joining you. It is another way to connect with your dog and if you, your dog, or both of you have trouble doing some of the poses, at least you tried a new way to be closer to your dog.

You might find that your dog wants to sit on the yoga mat as soon as you bring it into the room. If so, let her. Or call her over to sit on the mat if she doesn't take the first step. Then sit down next to her and slowly start doing some yoga moves. She may decide to stretch out and just watch while sitting with you. She may lose interest and head out of the room or decide to watch you from a distance but even that is a beginning. What poses you can actually do that incorporate your dog depends upon how big he or she is. If he is smaller, you can lift him up when you are doing, say, the Warrior Pose. If she is a larger dog, you can lean back toward her while you are doing the Camel Pose.

You can also gently stretch her legs (ask your vet first about the right ways to do this; you don't want to stretch anything in the wrong direction or pull too hard) and massage her legs and back or tummy. Even if you both just rest in a quiet state on the yoga mat, you will feel closer to each other and more connected. The relaxation will be good for both of you.

CHAPTER 7

THE THIRD STEP: RIGHT SPEECH

Black Dogs

Prejudice. People who are prejudiced often don't think that they are prejudiced. In their way of thinking, they simply believe what they believe they are justified in believing, even though the definition of prejudice is a "preconceived opinion that is not based on reason or actual experience." To the prejudiced person, there is no room for a different interpretation of their bias. Their opinion is their opinion and they are entitled to it. The real trouble starts when a prejudiced person acts on their prejudice. Then we have discrimination and that can really hurt.

"Be careful with your words. Once they are said,
they can be only forgiven, not forgotten."
—Carl Sandburg

The words we use can add to prejudice. Words are very powerful; what we say, we become. Humans use something like fifteen thousand words a day. Be measured in how you use them.

Believe it or not, some dogs—a lot of dogs, apparently—are overlooked for adoption simply because of their color. Black Dog Syndrome is prejudice and discrimination against dogs that are black, and it is a real issue. Why I cannot fully understand, since most of the dogs I have adopted have been all or mostly black. Dogs (and cats) who are black have a harder time getting adopted. Some say this is simply not true, while some question whether there actually is a bias against black dogs since more dogs and cats who are black enter the shelters so more are in the shelters waiting to be adopted, hence fewer are leaving the shelter simply because more of them enter the shelter in the first place.

But why are more black dogs brought to the shelter to begin with? It seems that a higher number of black dogs are euthanized. And how many black dogs are actually adopted versus the number that come into the shelters? How long does it take for a black dog to be adopted? Does it take more time—does the black dog spend more days at the shelter than do his lighter colored brothers and sisters?

When searching for a dog to adopt, often the photo of the pup and a very brief description is all the potential pet parent has in the way of information to make an adoption decision. Often, photos of black dogs don't show their eyes or other defining characteristics if the lighting is bad.

According to a *Los Angeles Times* article, Black Dog Syndrome is a reality for rescue groups and shelters across the US and involves black cats as well as black dogs. While there may just be more black pets in the rescue population in general, the problem is recognized by shelters across the country as a genuine obstacle that negatively influences the adoption of pets who happen to be black.[19]

Here is a random look at how having black fur has played out for dogs throughout history. In Japan and parts of the United Kingdom, black cats are considered to be good luck. Black dogs have usually had more sinister dogmas associated with them.

"If you meet the Black Dog once, it shall be for joy;
if twice, it shall be for sorrow; and the
third time shall bring death."
—W.H.C. Pynchon

Black dogs have been seen as evil in some legends and literature. Take for instance, the legendary Black Dog of the Hanging Hills in Connecticut. The tale has been around for more than a century and tells of a black dog who seems harmless and even friendly, following the traveler for a distance. All well and good, but if the same traveler sees this dog a second time, sorrow is sure to follow, and if the black shows himself to that traveler a third time, well, death in some form is just around the bend. So, thought W.H.C. Pynchon, a New York geologist, who, perhaps jumping to conclusions, found this to be so...for him and for fellow geologist Herbert Marshall. Anyway, when they were out and about in the area for research. After seeing the black dog for a third time Marshall slipped, fell, and died.[20]

Then there are the many variation on the malevolent Black Dog of the British Isles, a ghostly canine version seen as something to be feared, although a black dog can also be seen as a guardian.

And of course, there's the case of the poor fictitious Hound of the Baskervilles, which turned out to be not the threatening, glowing monster of a black dog it appeared, but just a large black dog painted to look fiery in the dark.

No need to beware the black dog. Instead, notice the regal, understated beauty of their coat—no need for a flashy splash of tan, brown, or white. The eyes of a black dog are spotlighted and if you look closely, you just may see the depths of the soul behinds those eyes. A black dog knows it takes a special person to notice their special place in the world.

Petfinder is an online database of animals needing to be adopted and is a directory of more than ten thousand animal shelters and adoption groups in the US, Canada, and Mexico. Petfinder conducted a survey that found that member shelter and rescue groups reported that while most pets are listed for about twelve weeks on the Petfinder site, animals seen as being less adoptable, for instance, black, senior, and special needs animals, languish nearly four times as long on their site.[21]

Regardless of whether Black Dog Syndrome is real, an article on VetStreet.com maintains this is a debatable point since, either way, it is a given that there just are more black dogs and cats looking for homes, and any extra promotion and publicity they can get is much needed.[22]

As more people have become aware of the possible prejudice against adopting black dogs and the large number of black dogs (and cats) in shelters, it may be that this is resulting or will result in more black dogs being adopted, hopefully.

Indiscriminate Breed Discrimination

Another area where people can be guilty of unjustness is when they are prejudiced against toward certain types or breeds of dogs based solely on the breed of the dog, not taking into account myriad other factors, such as how the person has or hasn't trained the dog.

Best Friends states on their website: "To keep people safe from dog-bite-related incidents, the best and most effective laws focus not on breed, but on behavior—both the pet owner's behavior and the dog's behavior."

A number of studies have shown that often it is the owner who is clueless about their dog's behavior and how to manage it. And not paying attention to the dog or the surroundings is another factor.

Spotlighting one breed of dog can give a false sense of security to a community. Studies of potentially preventable fatal dog bites from 2000-2009 gathered by organizations such as the National Canine Research Council have concluded that the breed of the dog was not a factor and that focusing on specific breeds in legislation is not effective. Instead many components should be included, such as the owner's failure to spay or neuter the dog, dog isolation, and owners with a known history of abusing their dogs. No matter how you may think you feel about bull terriers, these dogs are very popular and are part of many communities in the US.

There are many pit bull terrier-like puppies and dogs in shelters around the country, and they often face death instead of finding a home. There is also the issue of overbreeding pit bull terriers and a lack of spaying and neutering resulting in overpopulation.

A common problem around pit bulls is that they are often one of the most misidentified types of dogs. There is no real agreement about exactly what breed or breeds of dogs are pit bulls and what breeds are not.

It seems that no one can agree on what breed exactly constitutes the term "Pitbull" or "pit bull." The name is really an umbrella term and instead refers to a type of dog, rather than an actual breed. In reality, there are around thirty breeds that could fit under this very broad definition. Over the decades, different breeds have fallen on the outs with society. Doberman Pinschers were targeted in the 1970s; in the 1980s, it was German Shepherds; and the 1990s saw a prejudice against Rottweilers. Consider that the Labrador Retriever is banned in the Ukraine.

The American Kennel Club (AKC) has taken a stand against the banning of breeds and supports putting the responsibility where it belongs: with the human in the relationship. The AKC does not recognize "pit bulls," though some recognized AKC breeds, such as the American Staffordshire terrier, are often called pit bulls by people who don't know any better. The United Kennel Club (UKC) does recognize the American pit bull terrier as a breed.

When someone mentions "pit bulls," it could mean any of these breeds or combination of these breeds: American bulldog, American Staffordshire terrier, American pit bull terrier, Staffordshire bull terrier, and English bull terrier. But what's in a name becomes important when naming the breed becomes part of laws, ordinances, insurance policies, military housing and rental leases, and more.

Pit bulls were not always despised dogs. In the past, they were considered the embodiment of all that was admired in dogs. They were once referred to as the "nanny dog," and were considered

"America's Dog." One of the main characters in the classic show *The Little Rascals* was a pit bull, Pete, with the added black circle around one eye giving him a special signature look. Kids and this dog just went together. A pit bull named Sergeant Stubby served as the official mascot in the twenty-sixth division in World War I. He is probably the most decorated dog of that war and saved his troops from many surprise attacks.[23]

How did the pit bull go from being the nanny dog to such a vilified breed? Things changed in the 1980s with the revival of dog-fighting, and this is the breed that was associated with it. The pit bull became synonymous with a kind of macho style and the population mushroomed in some areas.

While the bite of a strong-jawed dog can inflict serious injury, statistics show that the dog that bites the most is likely to be a chihuahua, dachshund or Jack Russell terrier. Most dog bites involve dogs that are not spayed or neutered.[24]

As for that lethal dog bite, you are more likely to die from: contact with hornets, wasps or bees; an assault by a firearm; heart disease; cancer; chronic lower respiratory disease; a motor vehicle crash; or exposure to excessive natural heat—to name a few fatal possibilities—than from a dog bite.[25]

Be Aware of What You Say

Words can be very powerful and carry more weight than the speaker may have intended. Words can have consequences. Words can relay the truth, or they can hurl falsehoods and hurt. Always use words in a positive way. This is especially meaningful with your dog. Many trainers suggest going with positive reinforcement rather than scolding or using hard words. There are times when your adopted dog will need to hear a firm enforcement word like "No!" It will also be helpful to discover what may have been in his past that causes him to react in the way he does.

Be aware of what you say, but also be aware of how you say it. Shouting or angrily trying to get your dog to listen to what you are trying to train him to do will only make the dog fearful, confused, and even resistant. Charlotte is very sensitive to correction and to tone of voice. Telling her the command in a normal voice is usually sufficient to get the message across. Saying it loudly out of frustration doesn't work in the long run. All that happens is that the pet parent just gets more agitated while the dog looks on in bewilderment.

Working to dispel the prejudice against black dogs and certain breeds is one way to take another step on the path to Enlightenment. If you are thinking of adopting a dog, be aware that many dogs, both older canines and puppies, are often passed over because of their color. Consider purposefully adopting a black dog. And think about what you might have heard or have been saying about certain breeds of dogs. Do some research to find out the general characteristics of some of these breeds. Information about breeds can tell you certain characteristics in a general way, but each dog is different. Dogs for adoption are usually mixed breeds and no one can honestly tell you what breeds are in his or her makeup because the adopters just don't know. You can bring a trusted dog trainer with you or go on your own instincts when you meet the dog and try to find the dog who just seems like he or she should be a part of your family.

Even if taking these steps is unconsciously tied to making the right speech, the steps are still footfalls in the right direction. These efforts will lead to hopefully a more positive outcome for all dogs of every color and breed.

A Meditation on Your Dog and Her Characteristics

When you look at your dog, what do you notice first? Whether you and your dog are new friends or have been family members for a long time, there is always something new to discover about her or him. Sit quietly with your dog and notice a feature like her nose or her ears, for instance. How intricate the ears are (and how hard, sometimes, to keep clean). Trace the shape of an ear gently with your finger. Concentrate on how it feels and how it moves with your touch or with different sounds. Your dog is a furry bundle of complex elements and noticing how beautiful they are—and how vital they are to her—can help further strengthen the bond you have with each other.

CHAPTER 8

THE FOURTH STEP: THE RIGHT ACTION

Actions speak louder than words. If you're gonna talk the talk, you gotta walk the walk; the shortest answer is doing; well done is better than well said (that last one comes from Benjamin Franklin). There is no shortage of aphorisms about saying versus doing. Words and actions. Sometimes you need to have words first to prepare for the actions. The danger is when we keep using words when we should have switched over to action, to doing something about the problem we have been talking about. And are still talking about. And will still be talking about.

Say you are ready to put your words into action. How do you know where to start? And how do you know if your actions are the right actions?

Charlotte is a black miniature poodle/pharaoh hound/Belgian shepherd mix with some other exotic breeds in the blend. Now that she is a little older, the Shepherd is starting to show in the way she sits, quietly and stoically waiting for her human to turn around and notice her. She can be the miniature poodle, big-little-dog at a moment's notice, though. At different times, the personalities and traits of the different breeds that make up Charlotte come to the surface—the super intelligent miniature poodle, the prey-driven pharaoh hound, the steadfast and loyal Belgian shepherd, and the bossy Wheaton terrier.

Charlotte is a skillful practitioner of the pharaoh hound bounce. When she can't contain her excitement for whatever reason, she performs the up and side to side hop. She also is a very accomplished food thief, an area in which this breed is known to excel. Her skin is stark white and very sensitive, another trait of the breed. And running in a strong zig-zagging fashion is another pharaoh hound specialty, all the better for them to hunt rabbits—something they are skilled at in Malta, their place of origin.

She is also very catlike. This may have something to do with some other unidentifiable breed in her or because she was fostered for a time with a cat. She will lick her thigh and then run her mouth over it like a feline cleaning herself. She will wait patiently as long as it takes for a chipmunk to stick its head out of its hole, so Charlotte can have the opportunity to pounce on the little trickster. She caught the tail of one once, but it escaped. That only served to cement Charlotte's obsession with getting another one

Charlotte also is able to figure a solution to a problem if the goal is important enough to her. She seems to have a more advanced way of processing solutions to problems. If she sees a ball or toy she wants inside of another dog's pen, she figures out a way to retrieve it, but only when no one is looking. A lot of dogs might sit there and whine or bark for the toy or bone. Charlotte seems to be able to do whatever it takes to expose the prize—but since she knows she is not supposed to have the other dog's bone (for instance, if the other bone is of the softer material for a puppy's teeth and Charlotte could bite off a piece and choke), she waits until the humans have left the room to go after the coveted prize. This happened on several occasions with a friend's puppy, Maverick, who comes to visit Charlotte regularly.

One time, I was helping to get the puppy ready to go outside for a walk when I noticed that Charlotte was gnawing on something. Charlotte can only have toys and bones only when she is being supervised. While some dogs might chew safely on these items, and others are known to suck on soft toys, Charlotte always ends up picking small pieces from the bone or toy and she tries to ingest it. This happened when she was given a black extra tough version of a dog toy that can be filled with treats. There were no treats in the toy, but she hadn't had it for more than a minute when she bit off a piece.

Clearly, someone has to watch Charlotte like a hawk when she's chewing on a bone or playing with a toy. What was she chewing on now? Her back was to us (of course). No one had given her anything to chew on; what did she have there between her paws, and how did she get it? It was a bone made specifically for puppies and the moderately soft, easy to chew material was something Charlotte definitely should not be gnawing on.

Somehow Charlotte had maneuvered the puppy bone from inside the puppy's indoor play pen and snuck out of the room when no one was looking. I grabbed it from her and saw she had already eaten some pieces. Luckily, they were smallish pieces, but she had swallowed them just the same. The worry whirl started up again. Were the pieces going to choke her? She seemed to be swallowing and breathing okay. Would they hurt her internally? The pieces were smaller than the little chunk of Kong Extreme she had consumed a couple of years back. We would just have to be more watchful than usual over the next couple of weeks. Keeping her from harm has to be a priority.

Be aware of what you do

Don't do wrong is what this translates to. Do no harm to people or to animals. This step goes hand in paw with raising an adopted dog. Caring for the adopted dog with love and protecting him or her from harm is a part of the responsibility of the human, and in return, the pet reinforces your steps toward Enlightenment.

Looking for a dog to adopt, whether a puppy or older dog, is different from shopping for one. If you insist that must have a dog that has specific traits of a breed, you can try to find a pup or older dog who is a mix and might have some of the traits of the breed you are looking for, according to the description on the rescue site. You might even be able to adopt the exact breed you are looking for since there are a variety of dogs in shelters looking for homes. There is even the AKC Rescue Network (https://www.akc.org/akc-rescue-network/) that provides names and locations of shelters who rescue specific breeds from Affenpinscher to Havanese to pharaoh hound to Yorkshire terrier. You can find a dog of a specific breed who very much needs a home.

There is always the hope that there is the right human for every dog. The human just must be as sure as they can that they have the capability, the means, and the ability to give the adopted dog from a hoarding or other damaging situation the kind of specialized care these dogs dog often need.

Celebrities Helping to Promote Rescue Dogs

Anyone who adopts a dog should be celebrated. And while we don't need the experiences of celebrities to validate our feelings for our adopted dogs, some people very much in the public eye are such true examples of showing love for their dogs that they warrant mentioning.

The British actor Tom Hardy, for instance. When his dog Woodstock died, Hardy posted an open letter to his beloved pet on Tumblr, and his words were picked up by many media outlets. Hardy had rescued Woodstock, (Woody Woodstock Yamaduki Hardy) when the pup was little more than ten weeks old. Hardy had come across the pup running around a Georgia highway while visiting the state for a film shoot. Here is a portion of his open letter:

> *I don't normally speak out about family and friends but this is an unusual circumstance. Woody affected so many people in his own right so with great respect to his autonomy and as a familiar friendly face to many of you, it is with great great sadness a heavy heart that I inform you that after a very hard and short 6 month battle with an aggressive polymyositis Woody passed away, two days ago. He was only Age 6. He was Far too young to leave us and We at home are devastated by his loss I am ultimately grateful for his loyal companionship and love and it is of some great comfort that he is no longer suffering. Above all I am completely gutted. the world for me was a better place with him in it and by my side.*

> *To the bestest friend ever. To me and to a family who loved him beyond words and whom he loved without doubt more than I have ever known. Woody was the bestest of journey companions we ever could dream of having. Our souls intertwined forever.*[26]

Another who supports the adoption of rescued animals is Amanda Seyfried, who has said she can't imagine her life without her Australian shepherd, Finn. Her standard acting contract stipulates that he be allowed on sets in the US. Seyfried and Finn were in a video for Best Friends Strut Your Mutt and the #9000StepsChallenge. Seyfried is a longtime supporter of Best Friends. Seyfried also maintains that she learned a lot about being a mother from having a dog. "He was my firstborn," she has said.

Actress Emmy Rossum is another supporter of rescued dogs in general and of Best Friends in particular. She and her adopted dog, Pepper, joined Best Friends Animal Society's Save Them All campaign. "The wonderful thing about adopting from a shelter or rescue organization is that you save two lives, the life of the dog or cat you adopt and then another who takes the open spot at the shelter," Rossum says on the Best Friends site. "There are so many amazing dogs and cats of all ages, sizes, temperaments and breeds looking for homes and when you make that connection with the right one, your life is changed for the better."[27]

Miranda Lambert and MuttNation

Miranda Lambert is a singer-songwriter who believes in helping animals whenever they are in need. She adopted her dog, Delilah, to be a companion while Miranda was on tour. In 2009 Miranda and her mother, Bev Lambert, founded the MuttNation Foundation to bring awareness to rescue animals and animal shelters. They wanted every dog to have the possibility of finding a safe and loving home.

According to their website, "The mission of MuttNation Foundation is to promote and facilitate the adoption of shelter pets, encourage spay and neuter for all pets, and educate the public about the importance and beneficial impact of these actions. Our objectives are accomplished through numerous initiatives which include: high profile adoption events, fundraising events, a national pet transport network, and providing financial support to carefully vetted shelters across the United States. MuttNation is also committed to providing monetary, hands on, and transport assistance during times of emergency and disaster."[28]

Rachael Ray is not just a celebrity chef; she is also a big promoter of rescuing shelter animals. Her Rachael's Rescue has raised not only awareness of the plight of animals who are languishing in shelters but also funds to help these animals.

"Through 2016, Rachael's Rescue has donated more than $14 million to pet charities and other organizations that do good for animals," says her website. "The funds are used for food, medical supplies, treatments, and more. Many more animals around the country need help, and through Rachael's Rescue, together we can make a difference in the lives of many four-legged friends."[29]

When celebrities pool their resources, animals benefit even more. As of June, 2018, Ray's foundation had donated $27.5 million[30] to rescue charities that are close to the heart of other celebrities like Miranda Lambert, Hillary Duff, and Carrie Underwood.

A former president has also given the nod to adopting rescue dogs. In 2016, George W. Bush and Laura Bush adopted a puppy after visiting the SPCA of the Texas Jan Rees-Jones Animal Care Center. They had gone there to express thanks for the terrific work they do. While there, they found a puppy to adopt, took him home, and named him Freddy. He gets along very well with their two cats.

As it says on the George W. Bush Presidential Center site:

> *"If you could use a little extra joy in your life, consider adopting a pet from an animal shelter or rescue group."*[31]

Former Vice President Joe Biden and his wife Jill adopted a German Shepherd from the Delaware Humane Association in November 2018. They had first fostered the pup, named Major, who was from a littler of pups that had to undergo lifesaving medical treatment when they were rescued. [32]

The Royal Family of Great Britain also walks the walk when it comes to recognizing and helping homeless pets. The Duchess of Cornwall is very focused on animal welfare. In addition to caring for two rescue dogs, Camilla is a patron of the Battersea Dogs and Cats Home, a rescue organization that has been around since 1860 and cares for more than seven thousand animals every year.

These are just a few of the people in the spotlight who have used their prominence to help animals in need. Anything that enables more people to see what they can do to help animals in need is a very good thing.

Many other people in the spotlight contribute to bringing awareness of homeless pets and the shelters that care for them while they wait for their forever homes. The television program *Save Our Shelter* travels to different animal shelters around North America to help the people involved in animal rescue. The team includes pet rescue expert Rocky Kanaka, builder Rob North, design expert Emmanuel Belliveau, philanthropy experts Cathy Bissell and Aimee Gilbreath, and a host of corporate sponsors. They all work together to help animal rescuers with pressing problems facing their shelters. During each segment, one pet is spotlighted for adoption by the end of the episode, but all the animals benefit with the improvements made to the shelters. *Save Our Shelter* wants to be a "movement to enhance the lives of homeless pets in North America,"[33] and they are doing this while providing viewers with an entertaining show and information about the seriousness of the problem of homeless pets.

Being a celebrity or someone in the public spotlight doesn't give you special powers, but it does give you special opportunities to draw attention to problems like homeless animals looking for good homes. It is the right thing to do. Helping to raise awareness of the animals in shelters can only increase the chances that more will be adopted. When an animal is adopted from a shelter, as the saying goes, two animals are helped. The animal adopted and the homeless one who will be able to be taken in by the shelter and then also be adopted. And so it goes—until they all have a home.

Journaling Toward Acceptance

Now might be a good time to take twenty minutes or so to put some thoughts in a journal, or at least to think about how you have been frustrated because you are so sure that what you have been thinking, saying, or doing around any number of things has been wrong. Are you being too hard on yourself?

In what ways are these actions or behaviors wrong? Is this really so? What about all of the things you have been thinking, saying and doing that are right? You can take the self-improvement thing too far, you know. While we all can benefit from trying to better ourselves, balancing that goal with recognizing what we already have positive going for us is healthy. Write down in the journal as many things that you have done and are doing right. It just might help you to trust that you know what and who you are. Most of the time, anyway. And that's a start.

CHAPTER 9

THE FIFTH STEP: THE RIGHT LIVELIHOOD

We all need to make a living. Having a job or work that enables us to have the funds for the basics—food, shelter, clothing—for us and those we care for. How do we know if the work we are doing is the best we could be doing, if there wasn't some other means that would allow us to earn a living that might not only be better for our mental and physical health, but would allow us to have more time for those we love?

Earn a living by doing work that helps rather than hurts. Don't do work that goes against Right Speech or Right Action. This step also involves being responsible for your adopted dog and being there for other animals in shelters.

When you have any kind of pet, it is important to have a solid relationship with your vet. Your pet cannot tell you what is wrong, and while it might be possible to guess some ailments or injuries, for example, if the dog is limping, or walking differently, or behaving differently, the reality is that dogs do not show pain the way humans do.

There are times when you will need help getting your dog to take medication prescribed by your vet. Sometimes medication can have a particularly nasty taste to a dog, such as the often-prescribed for giardia metronidazole. It is not easy to convince a dog to take it. Wrapping it inside a slice of an organic turkey can work, and the pill can also be crushed and "hidden" in the mix of regular dog food. But in the case of Alice, who was a very picky and poor eater, she always knew when there was something different in the food, which she didn't devour with any relish in the first place. If the medicine is in a powder form it can be crushed and mixing it with peanut butter (if your dog tolerates peanut butter) is a way to hide the medicine. There are so-called pill pockets for this purpose, but they do not always work. If the dog is suspicious in the first place, she will not be fooled by something that might look like a treat but is actually hiding a foul-tasting medicine. I remember the many times I felt a sinking feeling upon seeing and hearing the uneaten and now-mangled pill plonking to the floor, either with or without the pill case or at times, the turkey slice. This meant picking up the now slobbery and slippery pill and trying again. Sometimes, persistence paid off. Other times, it did not. Then a call to the vet was in order to find out what to do about a missed dose or a partially consumed dose of whatever medicine the dog refused to consume. Give it at the time of the next dose? Double up on the next dose? Keep trying? Having a vet who is understanding can go a long way to helping the pet parent care for their pet. While there are many ideas online for getting a dog to take his medicine, the truth is that they do not always work.

Determining that something might be wrong with your dog in the first place can be a challenge. It is well documented that dogs instinctively make sure any pain they are feeling remains undetected by others. It is the law of the wild. To show any weakness means they are at risk of becoming attacked. It is down to survival of the fittest. So, having a good relationship with your vet and trusting that he or she will go the extra mile to make sure the animal is not masking an affliction, injury, or illness, is vital. Regular vet checkups and pet health insurance is what many vets recommend.

Being a vet is certainly an admirable livelihood. Most veterinarians choose this profession because of their genuine love of animals, as well as having the necessary science acumen, of course. Earning a living by keeping animals healthy and seeing to their medical needs and being patient and understanding with pet parents requires the talent and ability to work with both the animals and their human counterparts.

On the side of the animals, it is usually the case that the dog or cat or other species has no interest in being at the vet's office or touched and prodded by a vet or a vet tech. Dealing with an actively reluctant patient who is equipped with sharp claws, teeth, and the aptitude to use them, requires great skill on the part of the medical person. And then there are the pet parents. Let's face it: at times we can be overbearing and very adept at using the internet to find out what might be going on with our fur babies and what can be done about it, as we should. But there are also people who are neglectful of their pets. The vet has to deal with all manner of pet parents and have the skills to deal with both the over-involved and the under-involved pet person. Since the animal cannot tell the vet the symptoms he or she is experiencing, that means it is up to the doctor and the pet parent to figure what is going on.

A dog's instinct to hide their pain or injury is also a factor. The initial exam often involves a lot of detective work in order to know if any medical tests should be ordered and if so, which ones. As with humans, blood work can tell a lot about what is going on with overall health.

Being a veterinarian is being a part of a profession that demands long hours and many expensive years of medical schooling. Being a vet can be very stressful and frustrating and for many, it can be very hard to cope with their work. Veterinarians can suffer from compassion exhaustion as well as physical and mental exhaustion. Vets are likely to see animals suffering from many different illness and injuries, some of which cannot be treated successfully. End of life care is also a part of the veterinarian job. According to one estimate, veterinarians experience death five times more often than do doctors for humans.

They often work long hours during the week, on weekends, and on holidays. Added to this is the sad fact that for many pet owners, the money available to spend on their pet's care is optional. Too often the pet parent delays getting treatment for their animal because of the cost and the animal's condition only gets more serious and more expensive to treat. Having pet health insurance can ease much of the financial burden but a lot of pet parents simply don't have it. Health insurance for pets is an added expense, but some policies that cover at least a portion of the cost of treating a pet for injuries sustained from an accident start at around twelve dollars a month; if treatment for an unexpected illness and accidents is added, the monthly cost is about twice that.

Most veterinarians work in a solo practice, which usually means they can set their own hours. But if they set up practice in an area where there are not enough vets for the number of animals in the community, the vet may find they need to work longer hours—much longer hours. And if the vet is part of a practice that has emergency hours (a godsend to pet parents) there will be required overnight shifts expected in addition to the vet's regular hours.[34]

Let's not overlook the important role the veterinarian technicians—vet techs—play in the care of our pets. Being a vet tech is a demanding job that requires skilled interaction with both the pet and the pet parents. Not unlike the veterinarian, the tech must navigate between the animal who cannot speak for himself and may be fearful or aggressive and the pet parent who may be helpful or not. Vet techs are also on the front lines when it comes to getting injured by an animal and also have to deal with a lot of death.

A 2014 CDC study showed that one of six practicing veterinarians has considered suicide. They often feel hopeless because so many of their patients die. Other studies confirm this. Veterinarians seem to be at a higher risk of suicide in contrast to people in general. There is not enough data about US veterinarians and suicide and how their work might play a part. Two summits about this issue came to the conclusion that more research is required to make determinations about veterinarians and mental health issues.[35]

Rebecca Johnson of ReCHAI has some thoughts about this. She doesn't think any increased suicide risk has to do with the relationship they have with animals alone. She has heard veterinarians tell her during national presentations to veterinary groups that there is death all the time. But number one, they come out of school with enormous amount of debt, and on top of that, they're dealing with death all the time. "There are those who would argue that if you continually have to euthanize animals," she said, "that that is very distressing. I think there's a combination of factors."

There can be compassion fatigue, especially with caring for animals in shelters. With animals being cared for in a shelter, the veterinarians, technicians, and other staff taking care of them must deal with the urgency of the situation in the shelter. There are always more animals that need to find shelter so that they hopefully can be adopted into good homes. Unfortunately, euthanasia is still a factor in many animal shelters.

American Humane has promoted the safety and welfare of animals for more than a hundred years. American Humane is behind the "No Animals Were Harmed" program in Hollywood, farm and conservation, and animal rescue.

American Humane, whose mission is to ensure the safety, welfare and well-being of animals, keeps track of the number of dogs and cats that are euthanized, but their most recent statistics are from 1997 and the numbers are from only the one thousand shelters that responded to the survey. In 1997, about sixty-four percent of the total number of animals that came into shelters were euthanized—approximately 2.7 million animals in those one thousand shelters alone. This may have been done for reasons of overcrowding, or the animals might have been ill, aggressive, injured, or in pain. Fifty-six percent of dogs and seventy-one percent of cats taken into animal shelters are euthanized.[36]

Because euthanasia is still a regular occurrence in animal shelters, it can be stressful for the staff working with the animals. They can get attached to the animals and do their very best to make sure that they're healthy, and that they have a good chance of being adopted.

It stands to reason that compassion fatigue can play a role for shelter animals caregivers as well.

I asked a veterinarian about the CDC study, and she said, "Indeed, there are commonalities of experience and common characteristics of vets that do make them more prone to severe mental and emotional anguish. It is a problem that we should be looking to address." This vet has personally discovered aspects of the profession that she believes have helped her kept her from going down this road. "Specifically, self-employment and maintaining work-life balance," she explained, "believing in and adhering to strict professional conduct, really listening to what owners are saying and asking of me, and, importantly, addressing what I believe the animals need from me."

Certainly, this is a genuine concern. A study found that veterinarians in Britain seem to be four times likelier to commit suicide than the general public. Research found that the long hours and very demanding work environment can be very hard to manage. There is additionally a lack of support from managers while at the same time, the humans in the client equation expect impeccable care with positive outcomes. This study also found that many veterinarians have a solo practice, so they have even less professional support and can feel alone without social connections, which can make them more susceptible to depression and suicide.[37]

Maybe it is time for pet parents to show their appreciation for the work their pet's doctors and technicians do. We need to temper the cost of the treatment and any frustration we may have with reaching a final and correct diagnosis for health issues our pets may have with the realization that our pets' doctors and veterinary technicians face these problems and more every working day.

Anyone who works caring for animals is doing commendable work—work that helps animals and helps the humans who are responsible for them is indispensable. It can be very stressful work most of the time. But when the vets, the techs, and assistants all work together with the pet and the pet parent, it is a confluence of nurture, treatment, and love that ensures the best is being done for your pet, a treasured member of your family.

Appreciating Your Dog

Live in the present. And be aware of yourself in present. Be mindful of your body, your emotions, in an objective way. Judging is not an option. Sharing the present moment with your adopted dog helps ensure that you will stay right there in the moment with her or him.

Perhaps take some time to think quietly about your dog and his or her health at the moment. Meditate on only the present and how you feel when you are thinking about your dog. He or she doesn't have to be with you. Visualize her face; hear her sounds; imagine her energy. Don't think about what you must do for her. Just think about how you feel when you think about her and concentrate on the feeling. Don't think about how you think you should feel; just how you feel. Focus on the joy you experience just by thinking of your canine family member and the happiness she freely gives to you. After all, what we are looking for is right there inside of us. We just need to see that it is there.

CHAPTER 10

THE SIXTH STEP: THE RIGHT EFFORT

Finding the Rescue Who Will Rescue You

You mean to do better. You really do. You want to be true to your principles and put your altruistic words into meaningful action. You have given a lot of thought to the animals in shelters in your area and if there is anything you can do to help. Maybe you are not quite ready to commit to adopting a dog, but as long as where you live permits dogs, you might consider fostering a dog. This involves living with and caring for a puppy or older dog and helping him or her get adjusted to living in a home environment. Fostering a dog requires the same amount of time and care an adopted dog needs, but fostering is temporary.

If you are not able to foster a puppy or dog, there are other ways to help animal shelters. Provide transportation for a dog (or dogs) from the shelter to various destinations or donate supplies. Shelters often have a wish list of items they need, ranging from dog food to office supplies. Shelters can always use more volunteers, so if you can donate your time, that is another option.

Making sure we are doing all that is possible to be proactive about our dog's health is part of the Right Effort step to Enlightenment. Even if taking these steps is subconsciously tied to making the right effort, the steps are still footfalls in the right direction. These efforts will hopefully lead to a more positive outcome for our dogs.

Now was the chance to take Charlotte's health, wellness, and longevity in hand. If there existed tools to ensure this as best as possible, I would make use of them. It was likely that her pet health insurance would not cover most of these wellness exams and tests. But what better way to spend money than on a truly special dog who is giving so much to me and my family? Given the choice of paying for comprehensive preventive medical tests for Charlotte or buying, say, a new coat, I would gladly choose Charlotte's needs. What happened with Alice is a tattoo on my heart that reaches deep into my soul.

"When I let go of who I am, I am able to become who I might be." A satiating slice from the Lao Tzu wisdom cake. The rescue dog doesn't care who you are or who you might be, not even who you once were. Many animal rescue organizations—such as Best Friends—make the right effort. The work that rescue groups do is a prime example of putting the needs of others ahead of our own.

Earn a living by doing work that helps rather than hurts. Don't do work that goes against Right Speech or Right Action. This step even goes as far as recommending some jobs to avoid, such as anything that can cause harm to others, animals or people. Being responsible for your adopted dog reinforces this principal. The dog will follow your lead and behave as you do.

When you think of organizations that look out for animals you think of the ASPCA, Best Friends, or the SPCA. The history of the prevention of animal cruelty goes back a number of years, and across the pond to the UK.

According to the Royal Society for the Prevention of Cruelty to Animals (RSPCA), there are about 8.5 million pet dogs in the UK. That number represents a myriad of different breeds and types of dog; they are all of different sizes and shapes, and each has their own unique personality.

The importance of protecting animals, including dogs, started to emerge and in 1824 the Society for the Prevention of Cruelty to Animals was founded in a London coffee shop. This was the world's first animal welfare organization. The "Royal" was added in 1840 after Queen Victoria allowed the organization to add the royal R. There had been royal patronage associated with the organization since 1837, but now the charity was officially royal.

When the SPCA was founded, they specialized in helping working animals, such as the ponies who labored in the mines, "pit ponies," as they were known. The World Wars brought change and the RSPCA worked to aid the millions of animals that served with the British, Commonwealth, and Allies.

Two years before the Society for the Prevention of Cruelty to Animals was founded, Martin's Act was passed. According to the RSPCA's website, this was the first animal welfare law. Martin's Act outlawed the "the cruel and improper treatment of cattle." In 1835, Pease's Act extended this law to dogs and other domestic animals.[38]

As the RSPCA says on its website, the most significant change since the founding of the organization has been a shift in attitude about animals. The UK may be considered a land of animal lovers today, but this wasn't always the case. In the beginning, changing people's attitudes toward animals from thinking of them simply as commodities to viewing them as sentient creatures proved difficult.[39]

About every thirty seconds on average, someone in England and Wales dials the RSPCA twenty-four-hour cruelty line for assistance. The RSPCA received 1,153,744 such calls in 2016.[40]

The Animal Welfare Act became law in England and Wales in 2007. Pronounced the most important example of animal welfare regulation in almost a century, the Animal Welfare Act puts the legal responsibility squarely on the owners and keepers of animals to take care of them appropriately.

In 1866, a few decades after the UK SPCA was founded, New York City's Henry Bergh, a diplomat and philanthropist, founded the American Society for the Prevention of Cruelty to Animals. There is a connection of the ASPCA's founding to the Royal Society of the Prevention of Cruelty to Animals.

Bergh had been a diplomat to the court of Czar Alexander II in Russia. He didn't like the way the peasants treated their horses, often beating them. The cruelty left an impression on him that he could not forget, and he resolved to do something about it. After he stopped in at the Royal Society for the Prevention of Cruelty to Animals on his way back to the United States, he realized that here was a concrete way to not only educate people about the right way to treat animals, but to have the power to arrest and prosecute anyone who flouted laws against animal cruelty.

Bergh returned to New York and at a February 8, 1866, meeting at Clinton Hall, he argued that protecting animals was an issue that crossed party lines and class boundaries and that it was purely a matter of conscience. His speech moved quite a few dignitaries to sign his "Declaration of the Rights of Animals."[41]

The New York State legislature passed the charter that incorporated the American Society Against the Cruelty to Animals on April 10, 1866. But there had to be teeth behind the charter, and the first meaningful anti-cruelty regulation in the United States was passed nine days later. The ASPCA could officially investigate complaints of animal cruelty and arrest those responsible for being cruel to animals. The ASPCA swiftly became the prototype for more than twenty-five other humane organizations in the United States and Canada.[42]

Another organization that is at the forefront of animal welfare is Best Friends. In the 1980s, at a time when the usual way for shelters across the US to deal with unwanted pets was to euthanize them and about seventeen million died every year—with the sick, older, and difficult ones the first to go—some friends got together and worked to start bringing a number of those so-called difficult pets to a place that was safe and secure. That group of friends eventually became Best Friends Animal Society.[43] Based in Kanab, Utah, Best Friends has a mission to bring about a time when there are "No more homeless pets." Best Friends' goal, and their slogan, is to "Save them all."

In 1984 the founders of Best Friends started building a no-kill animal shelter in Kanab, today the headquarters of the largest no-kill animal shelter in the United States. But what does "Save them all" mean? Best Friends aims to ensure that someday, no animals in shelters will have a death sentence. The organization also works to try to reduce the number of animals who end up in shelters in the first place. And they collaborate with other rescue groups, government, and individuals who are interested in supporting their mission.

Working together with others ensures that even more animals will be rescued.[44]

In 2012, Best Friends started No-Kill Los Angeles (NKLA), with the plan to make L.A. a no-kill city. That same year, more than sixty communities across the US in the United States reached a ninety percent no-kill save rate. By 2013, there was a fifty percent reduction in the number of animals being killed in shelters, putting the goal of making L.A. a no-kill city by 2017 within reach. Since its inception, the total save rate in the city of Los Angeles rose from 57.7 percent in 2011 to 85.96 percent in 2017. The save rate for dogs has improved from 71.3 percent in 2011 to 91.93 percent in 2017, exceeding the no kill goal of 90 percent.[45]

In 2014, Best Friends started No-Kill Utah (NKUT) with the goal of having Utah be no-kill by 2019.

Other animal rescue organizations came into being and grew over the years. One is Maddie's Fund. In 1999, PeopleSoft founder and CEO Dave Duffield named a foundation they had stated a few years earlier Maddie's Fund. The $300 million behind the fund was to support the no-kill choice. Maddie's Fund is named after the Duffields' Miniature Schnauzer, Maddie, who had been an inspiration to them during the challenging early days the PeopleSoft startup. Since then, Maddie's Fund has awarded more than $153 million in grants toward this goal, including shelter medicine education.

Maddie's Shelter Medicine Program is based at the University of Florida College of Veterinary Medicine and exists because of Maddie's Fund. This wonderful entity has as its mission to give veterinary students and practitioners the focused knowledge and skills to improve the medical and behavioral health of animals in shelters, for starters. They are doing this by—among other things—training veterinary students and practitioners to understand and work with homeless animals on the issues and challenges they face, from the shelter system as a whole to the plights of the individual animals in the shelters, including medical and behavioral concerns. Shelter medicine is a specialty and veterinarians and other professionals need to be able to treat it as such.[46]

In 2000, funded by Maddie's Fund, Best Friends No More Homeless Pets in Utah program began the first statewide no-kill campaign. More than one-hundred thousand animals were adopted and there a now twelve no-kill communities and the number is growing. There is a more than eighty-five percent save rate for dogs across the state.

Michael Arms, considered a pioneer in animal welfare, started Home 4 the Holidays, which turned the adoption of shelter animals into a national regular marketing event that led to the adoption of several million shelter pets. Mike was an adoptions consultant for national and international adoptions. He then became president of the Helen Woodward Animal Center in 1999.

The story of Michael Arms' commitment to shelter animals is an inspirational and powerful one and deserves to be shared here. Disturbingly, animal cruelty is not new. Mike was forever changed by something that happened to an animal. Mike had been a veteran with the ASPCA and the North Shore Animal League. He had just handed in his resignation at the ASPCA and during the remaining few days he was called to see to a dog that had been hit by a car. The dog that had been hit so hard that the crash broke the dog's back.

Three men at the scene wouldn't let him take the dog away; they were taking bets on how long it would stay alive. Mike was attacked by the men when he ignored them and tried to bring the injured pup into the ambulance. Mike was beaten very badly. Amazingly, the dog somehow crawled over to Mike and licked him until he was conscious. That was when Mike vowed that if God let him live, he would do everything he could to protect the forgotten and abused animals.[47]

This brings us to the Helen Woodward Animal Center, a private nonprofit organization in San Diego County, a nationally recognized shelter, which has been committed to the philosophy of people helping animals and animals helping people for more than forty years.

Mike eventually became president of the Helen Woodward Animal Center where he helped the center's programs evolve and the adoptions group doubled in number. This led to three times as many pet adoptions each month.

The Helen Woodward Animal Center understands the value of getting to people at a very young age to help them learn about the humane approach to caring for animals, and has reached out to a record number of young people. The center also helps other shelters through workshops and conferences that show how to successfully run a nonprofit organization. The idea is to apply the principles of running a business to running a nonprofit so that the nonprofits can more successfully help the people and pets they are intended to assist.[48]

Another program Mike started is the Remember Me Thursday global awareness campaign. This program asked people to light a real or virtual candle every year on the fourth Thursday of September to help disseminate awareness of shelter pets and to get the message "opt to adopt," out there in their network. Everything we can do to reduce the millions of homeless pets who are euthanized each year can have some impact. Social media has been a big factor in the growth and success of this initiative.

These stories of people giving of their time, and money if they have it, all center around a love for animals that is practical and concrete. There is no "If only" or "I wish." These are people who recognize that the way animals are treated needs to be changed and they are willing to do what it takes to be the force of that change. And their stories and actions all interconnect like the warp and woof (woof, indeed!) in a tapestry.

The Helen Woodward behind the Helen Woodward Animal Center has her own story. Helen Whittier Woodward, the center's founder, pledged to make the world a better place for both animals and people back in 1972. She has made good on her commitment. Helen always had dogs, and they always had a special place in her heart—and in her home. They were allowed to sit anywhere on the furniture and when somebody visited, they often couldn't find an available place to sit.

After Helen's two children were grown, she began looking into the idea of starting an animal shelter. Together with a longtime friend who was experienced in training and caring for animals, Helen bought an old farm and established the San Dieguito Animal Care and Education Center. Helen's vision was to create a special place where "people help animals and animals help people."[49]

"It isn't necessary that you leave home.
Sit at your desk and listen. Don't even listen,
just wait. Don't wait, be still and alone."
—Kafka

Get started. Do something. Effort, or the striving toward, is what motivates all of the steps in the Eightfold Path. If there is no effort, nothing will get done. If there is effort, things will get done, so be sure to begin projects, work, or tasks that will lead to good. Your dog understands this. Here is where the adopted dog can help, as well. Making sure you are doing all that is possible to be proactive about your rescue dog's health is also part of the Right Effort step to Enlightenment.

Being still. Being quiet. Some people get to this point using mediation. But aren't there many types of meditation? Following your instincts can help you can reach a place of deep concentration, even if you're just sitting with your dog quietly and listening to him breathe or following her small movements while she is at rest.

Follow each breath your dog takes with your eyes and your hand. See how even though she is completely still, she is not motionless. Her body is alive with so many small movements and sounds, twitches, and sighs. Even when she has a nightmare and breaks her own sleeping spell momentarily, you are a part of her, comforting her, and watch how she wakes enough to acknowledge that whatever happened in her frightening dream did not really take place and you are there to comfort her and be there for her.

See as she returns to the regenerative powers of rest and sleep. So much energy in the limbs and attentiveness in the ears, eyes and especially the nose, all quieted and calm. Feel her foot, the pads and tufts of hair in between.

I remember when my dog was a puppy and how smooth her foot pads were. Years of using them to go in every direction and take care of her important business have toughened the foot pads, but they are still soft enough.

Touching the dog's feet is important so that the dog feels comfortable with you handling this very sensitive and important part of her body. A dog's feet support her, of course, when she stands, walks, and runs, but they also are where most of the dog's sweat glands are located. Feel the dog's foot, how the pads are supple and yet strong at the same time.

A Dog's Paws
—Sam Logue

I would touch the bottom of his feet when I was small
Amazed by the feeling
Three coarse, pad-like structures
Each one protecting him, from snow, from rain, from rock, from dirt
I looked at his paws and wondered why I didn't have such a thing
My feet would bleed on rock, slip on snow
Twenty years later, I still remember—
The feet of my best friend.

A dog's nails also are a vital part of her foot. Trimming your dog's nails is an important part of their care. With a rescue dog, there may be some uncertainty on the part of the dog about having their nails trimmed. Some people leave this grooming chore to their vet. Others do the nail trimming themselves at home. It is a good idea to let the dog, whether grown or a puppy, get used to whatever tool you will use to trim her nails. Having experience with a wide array of options as far as what tool to use for trimming the dog's nails, the person at Charlotte's home who regularly trims her nails has settled on a professional nail grinder, a Dremel. This tool, a regular grinder without the protective plastic cap that has an opening to put the dog's nail through, does the job easily.

Charlotte was naturally afraid of it at first, but after a while with encouragement and a reward of a treat in the beginning, she now accepts that she needs to have this done.

Some find nail clippers the right tool for nail trimming, but Charlotte is a black dog and her nails are black, making it hard—if not impossible—to see the "quick."

You can click on countless blogs and websites and YouTube videos about the quick and how to avoid cutting it, but when you come right down to it, avoiding using anything that cuts when you are trimming nails with a nearly invisible quick, the Dremel is the way to go. Once the dog experiences having the quick of her nail cut, it can be a long and difficult road back to where she will even let you or anyone come near her nails again.

This happened with Charlotte. At one vet office there were two technicians, one holding Charlotte and the other wielding the clippers, and there was still a problem with clipping the dog's nails. Charlotte's quick was cut, causing her a lot of pain, and she jerked away; it was months before she was comfortable having her nails trimmed again.

What about that claw higher up on the back of their legs? Known as dew claws, do they serve a purpose other than needing to be trimmed? The answer is yes.

Because of where the dew claws are positioned on the dog's leg, this claw doesn't come in regular contact with hard surfaces that help to file the other claws as the animal walks along.

The number of dew claws that a dog has varies from breed to breed and from dog to dog. But the dew claw can serve a purpose for the dog. Some use it to help hold a bone as they are chewing. Some dogs make use of these extra claws to help them climb trees to "tree" their prey—the Jack Russell terrier and the Catahoula, for instance.

But regardless of how your dog uses her dew claws, they need to be trimmed and it is a bit harder to get a grip on these claws because they seem less fixed in in place and move about when trimming them. All of this must be done with the dog's cooperation and trust that you know what you are doing. You can't pretend with a dog. Dogs always know when we are faking it. If you are afraid you will hurt the dog when you approach her with whichever trimming tool you have in your hand, she will know it and run for the best (human-inaccessible) hiding place she can find.

It is probably best to set aside this grooming chore for the time being and come back to it another time. But do come back to it and try again.

One more grooming procedure you can do at home with your dog—and become closer in the process—is brushing your dog's teeth. Brushing your dog's teeth is something to get the dog in tune with as soon as possible. This good hygiene may help ward off dental disease but even it does not, your dog's breath will be fresher and he or she will get comfortable having their teeth cleaned. Some dogs even look forward to it!

Another way to come to terms with what you are doing and not doing is through journaling. Write down all the ways you are working toward making a good effort. Sure, you may not be doing all you think you should be doing, but the point is not to get stuck in the "I should" or "Why didn't I" kind of thinking. Instead of mentally wringing your hands, write about what you are doing now. You will find it easier to allow for the shortcomings of others when you have embraced your own. Leave room for what you will do in the future and trust yourself to know that when you do any or all of these things, it will be the right time to do them.

CHAPTER 11

THE SEVENTH STEP: THE RIGHT MINDFULNESS

It isn't ever easy to see things as they really are, to be objective and take off the filters of subjectivity that color our perceptions. To see clearly, to understand completely. If we can't see things as they really are, then how can we even begin to accept them? And if we can't accept them, how can we release them and the hold we allow them to have over us?

Live in the present. Be aware of yourself in present. Be mindful of your body, your emotions, in an objective way. Judging is not an option. Sharing the present moment with your adopted dog helps ensure that you will stay right there in the moment with her or him.

With George, a dog we had adopted when he was a puppy years before we adopted Alice, things went a bit differently than they later went with Alice. Many years before, George had been to an animal emergency hospital with what turned out to be a life-threatening rupture of his bladder. He had decided to jump from one landing to another and missed, and the subsequent injury became obvious shortly after. While paying the bill at the animal hospital, I noticed a line for people who had pet health insurance. I had not known about pet health insurance and in the rush of paying the bill and waiting apprehensively to make sure George's surgery was successful and that he would be okay, I forgot about it.

George had lived a pretty long life and when he was about twelve years old, he started acting like something was wrong. We brought him to a 24/7 animal hospital and after a long examination, the vet there told us that George had hemangiosarcoma—heart cancer. It seems that this type of tumor begins in the blood vessels in the heart and is a very common kind of cardiac tumor in dogs. It can start in the heart or it could have metastasized to the heart from another location. It can go unnoticed until it is too late. When the tumor becomes too large, it will burst, and this causes internal bleeding that can be fatal to the dog."[50]

So, this was how it was going to be for him. This is where understanding that we never really have control of things outside ourselves can help. It bears repeating: Zen and the rescue dog can go hand-in-hand, but the road to peace is not a straight line and there are many detours along the way. And there really is no end to the road, anyway.

"You cannot travel the path until you have become the path itself."
"No one saves us but ourselves. No one can and no one may. We ourselves must walk the path.
—Buddha

When some people are faced with something so unthinkable and enormously devastating that it is hard to go on, after some time they are able to turn that tragedy into a new purpose of helping others.

Be Aware That You Are Aware

This is how K9s for Warriors came to be. Based in Jacksonville, Florida, K9s for Warriors is an organization that saves warriors and rescues dogs. And the dogs in turn work with the warriors. Some of the warriors are veterans while some are still on active duty and all suffer from Post-Traumatic Stress Disorder (PTSD) or traumatic brain injury (TBI) which are life-threatening disorders.

Founder Shari Duval had experience working as a volunteer for veterans' charities in her area and she was impressed by the remarkable work they were able to do for military veterans. She wanted to find a way to help those who had disorders that were not immediately obvious, such as PTSD.

Brett Simon, her son, a veteran K9 police officer and private investigator, had worked as a contractor for the army as a bomb dog handler and after two tours in Iraq he came back home with PTSD. Brett and his dog Hugo were very successful on their missions, finding multiple inventories of weapons and explosives.

Shari and her family wanted to do something meaningful not only for Brett, but for others. They began developing the plan to establish K9s for Warriors to help post-9/11 veterans diagnosed with PTSD, TBI, and/or Military Sexual Trauma (MST). They researched canine assistance for PTSD and they concluded that starting a non-profit organization to train and give service canines to assist our warriors' efforts to return to civilian life with dignity and independence was what to do, and Shari she founded K9s for Warriors. Properly trained service canines are a viable option for recovery from PTSD and TBI.

Brett oversees Canine Operations at K9s for Warriors and created the training curriculum utilized by the program. Although he was challenged by this endeavor, he found himself wanting to go back to his true calling: working with dogs. Their goal is to expand their program to serve even more veterans. K9s for Warriors does not charge for their service.

About ninety percent of the dogs in the program are rescue dogs or from shelters. K9s for Warriors opened in 2010 and started with dogs two years and younger for the first couple years, since they didn't have the personnel needed for training puppies at first.

Eventually they started training dogs that were a little younger. The aim is for the dog to have a long life working with the veteran so the younger the dog is when he or she starts training with the vet, the better the dog will be able to help the vet for as long as possible.

Then as they and their staff grew, and the program developed, they were able to start taking younger dogs; by 2015, they were able to take on puppies. They started out with about four or five puppies and today they have more than forty puppies in training. Volunteers handle puppy raising, coming in at least twice a month to go through formal puppy classes. Outside of that, they work with the pups on day-to-day life tasks—taking the dog to work, the grocery store, and exposing them to all the sights and sounds so that the pups can grow learning how to work with their handlers.

The puppies stay in their training cycle for one year, still adhering to the same two times a month coming in for training. They can always come into the center at any time to get extra assistance if needed.

To find rescue dogs for the program, K9s for Warriors reaches out to shelters and rescues all around the state of Florida and a few adjacent states. They also have contacts in other states, including Tennessee, North Carolina, and Indiana who know the requirements they are looking for in a dog. When the dogs are medically cleared, they are transported in air-conditioned vans by a pet transport company to the campus in Florida. They opt for the air-conditioned vans, so they can be sure that the dogs aren't being put in the bottom of a cargo plane. It can cost from two thousand dollars to three thousand dollars to get the dogs to the campus.

The cost to train and place the dogs is about twenty thousand dollars. By raising and training their puppies, they have been able to bring down the cost a bit. Some larger providers of service dogs have 260 placements a year and have to manage all of the upfront costs like making sure the pup is up to date on all veterinarian care and shots, raising the puppy, training, and so on. Getting the dogs after they have been vetted and trained in the basics saves some of that expense, and since volunteers help train the puppies, it helps keep some of the costs down. Corporate sponsorships also help. Adopting older dogs saves twelve to eighteen months of extra costs that come along with raising the dog from a puppy.

Training, Training, and More Training

Brett has an all-encompassing dog training style that integrates positive reinforcement, classical conditioning, operant conditioning, and marker training. He believes that dogs need five elements for trainability: timing, consistency, praise, correction, and motivation. K9s for Warriors works with each of the dogs for six months.

Commands Service Dogs Need to Know

Stand: The dog comes out in front of the handler length-wise to create a barrier between themselves and whatever the stimulus is. This helps handlers to be able to go into a large, crowded room, and for a time the dog can be out there as a kind of a social buffer.

Then if the handler is feeling comfortable in the situation, he or she tells the dog to…

Heel: Pretty much the same as what it means to all dog-walkers, bringing the dog back to the side position, sitting right next to the handler.

Cover: The dog turns and faces the other direction and allows the handler to know that somebody is approaching either by a tail wag or leaning into them, so that the startle response is gone.

My lap: The command for anxiety—it is really a pressure therapy—where the dog actually puts half of his body up on the handler's legs, chest, or shoulder area, so that the dog is coming up and interrupting the handler's anxious behavior. The dog helps by nudging, licking, and applying pressure to the handler, letting him or her know the dog is there to help him or her through.

Brace: This command helps the vet out of a chair, helps them off the ground from a seated position, and so on. Just about every one of the warriors in the program has knee and back problems from their career, especially those who were in combat, since in combat they are carrying anywhere from eighty-five to one hundred or more pounds every day on their back. No matter how in shape the person is, it takes a toll. Just the day-to-day routine wears the joints and everything down.

Those are the main commands that dogs need to master to finish the final skills test to qualify as service dogs instead of an emotional support or a therapy animal. Emotional support dogs are not specially trained, and they need a letter from a licensed therapist to qualify. Therapy dogs offer a calming presence to people in very stressful locations, such as hospitals.

That is how the program works for the dogs. As for the humans, for three weeks the veterans/warriors live at the facility where they train and learn to re-enter civilian life. This, too, is a careful, thoughtful, and thorough process.

Candidates fill out a twenty-seven-page application on the K9s for Warriors website. The main sections pertain to matching the dog and the warrior to find out the warrior's lifestyle and the warrior's disability. Is their disability more cognitive than it is physical? Do they work? Are they in school? This part of the application consists of questions like these. Then the staff look at the attributes of the dog in relation to the applicants. Some of the veterans in the program are still actively running and lifting weights and still doing many physical activities but, at the same time, they are suffering from PTSD and/or PTI that might throw off their balance a bit. But they are still very active, and the K9s for Warriors staff knows they need to match candidates like this with a dog with a little more energy and drive than others.

Then they look at the attributes that the handler is going to offer. They know the dogs and their temperaments, their quirks. If the dog is on the lazy side, works really slow, and does commands really slow, then he may go to somebody who is not as active and mobile.

The staff sits down before the classes arrive, already knowing which dogs are prepared, trained, and ready to be placed with a handler. Once the folks at K9s for Warriors know that the dogs are ready, they sit down and look over the applications with the dog trainers and select the right dog for each handler on a tentative basis.

Once the applicants come to the campus, there is another meeting with them. The handlers live on campus for twenty-one days during which time the instructors teach them how to best benefit from the dog.

On Monday there is an early meeting where the applicants are asked basically the same questions that are on the application. These have to do with lifestyle, school, work, and so on. Then there is a more in-depth, one-on-one conversation with them. The answers on an application can tell a lot about the applicant but meeting the person and having a discussion face-to-face can be more definitive.

It is very important to match the right dog with the right need. The conversation with the applicants is a way to ensure that the preliminary pairing is the right pairing. Then the staff goes back to the drawing board right after they have sat down with every one of the potential handlers and hash out everything again. They all sit down and they put the pieces of the puzzle together.

It takes probably another hour to two hours where a think tank of people, dog trainers, warrior trainers deliberate the matches. Warrior trainers are people who graduated the program and can come back and teach the other veterans what they have experienced, which allows for peer-to-peer mentoring. They take the new handlers out every day.

If one or two of the dogs don't turn out to be the right fit with a handler Brett says they have a dog or two who is ready for final training if something doesn't work, or if the assigned dog develops a medical problem. Dogs can get sick just like people, and this can happen the day the class is starting, so they always have a couple of dogs that are ready for training who could be placed in the class. This means that for a class of twelve veterans, they may have sixteen dogs who are ready for class. This ensures they can fill the gaps if needed.

The applicants arrive on a Sunday night and the next day, Monday, everything gets started. "We pair them that afternoon—we call it our dog day," said Brett. "They meet their service animals right after their lunch time, and then we spend time with them throughout that day making sure the match is correct."[51] And it doesn't end there, because throughout the entire program, K9s for Warriors is always looking at what's going on.

Brett's warrior training style uses lectures, demonstrations, discussions, and hands-on learning. The most effective way for warriors to heal their PTSD is by being active as their own advocates in their recovery process. Brett wrote the training manual that is used in the program and it includes policies, tools, definitions, veterinary information, recommendations, and laws. There are also rating scales and checklists, so the handler can keep track of the dog's progress with training, as well as how the dog is behaving.

And there is proof that service dogs really do help the veterans. K9s for Warriors was involved with a recent study by Purdue University that shows that service dogs are associated with lower PTSD symptoms among war veterans. The preliminary study was led by researchers in the Purdue University College of Veterinary Medicine and showed that overall symptoms of PTSD are lower among war veterans who have service dogs. The pilot study was co-funded by the Human Animal Bond Research Institute (HABRI) and Bayer Animal Health.

K9s for Warriors has been working with Purdue for about four years doing a collaborative study about the human-animal bond and what the actual meaning of the service animal to their handler is. Checking levels of cortisol, the stress hormone, using mouth swabs can provide clear evidence of physical changes that result from the veteran working with a service animal. What are the tasks that the handlers love the best, and what is helping them to feel better? There is quite a bit of research being done into this. The National Institute for Health is rerunning the data and after a peer review, the study will be formally presented with Purdue and the National Institute of Health.[52]

One of the studies that is being considered or implemented is having both the service dog and the veteran wear fitness tracker-like wristbands to study when a veteran is spiking. This would show whether there's a connection between a handler's anxiety levels and their service animal's anxiety levels.

K9s for Warriors is a small charity doing huge work and making a difference. "There's over three-hundred and fifty thousand veterans that suffer PTSD or Traumatic Brain Injury, Military Sexual Trauma," said Brett. "I would like to see us expand in the future to be able to reach out, help more veterans." They have a second campus under construction that will allow them to work with even more veterans.

Right now, the K9s for Warriors mission is for post-9/11 veterans. They have talked about growing in the future and still keep their core program to maintain their "little family" environment and not get too big. "We want to do it slowly, but to be able to reach out to other people. Maybe even first responders, or back to other conflicts that have happened."

That is their goal. It is a daunting goal because there are so many post 9/11 vets and that's what their mission statement started out as: to serve post-9/11 vets. But they are growing and expanding and see the need to extend their reach. "We have a lot of partners that we work with that do all eras of veterans," Brett said. "So, if they contact us we put them in touch with the right people for them to be able to receive the same benefit we could give them. It's just through another program. We're very collaborative here." They are, and we are all the better for it.

Another organization dedicated to matching service dogs with people who need them is Tower of Hope. Tower of Hope was started in 2006 as a way for founder Cathy Carilli to help honor her husband, Tom Sinton, in a way that would help others. Tom was killed in the September 11, 2001 attacks on the World Trade Center in

New York City, as were the nearly three thousand other people just going about their daily business that bright, sunny morning.

Cathy wanted Tower of Hope to play a major role in helping anyone seriously wounded in the subsequent war in Iraq and Afghanistan to be able to have lives that were more independent thanks to service dogs.

Tom had a Labrador growing up, and he was fond of all animals. Cats, dogs, rabbits—they always put a smile on his face. After Tom was killed, Cathy found that her pets were very comforting. Whenever she would cry and the only thing she could hear was her racing heart beating, her pets would sit on her lap or just be near her. "I would pet them and eventually it would calm me down," she said. So, she wondered if animals could help other people who were going through trauma. "Thinking about that and doing some research on organizations that involved animals, I came across facilities that train service dogs, and from there recognizing that unlike Vietnam, a lot of the veterans were coming home disabled, where they would have died in the battle field." So, from there Cathy became determined to help veterans. Tower of Hope does help others with disabilities, but the primary focus is to get dogs out to the veterans because it takes quite a long time to train the dogs.

Training

The most efficient way to do this work is to partner with Assistance Dogs International (ADI) accredited facilities and from them, concentrate on veterans. (ADI is a worldwide coalition of nonprofit programs that train and place assistance dogs. They are considered the leading authority when it comes to assistance dogs.) "So, we are doing that by spreading the money out to several facilities that can focus on veterans," Cathy said. "We felt that that was the best way to get more dogs to more veterans." Tower of Hope works with ADI accredited groups because over time a lot of organizations started popping up, and they wanted to make sure they had that accreditation not only because of the requirements of the facility, but also the requirement to follow up and retrain the dogs if need be. It is extremely important that the dog be as sharp as he or she can possibly be. Dogs are reassessed every two years, but the ADI facilities always have their door open to their recipients, so they can touch base with the trainer or somebody else from the organization.

When a veteran applies for a service dog, they go through a process that includes a psychological evaluation, a physical evaluation, and a test of their ability to remember and use commands. This can be very hard with people who have traumatic brain injury because sometimes it's a challenge for them to remember the commands.

During the process, there are evaluations by specially trained professionals because they don't want to put a dog in a position where there might be alcohol and violence in a house. Once the applicant enters the program, "There are times where a dog comes in and it will go right to an individual," said Cathy, "and then that dog is trained more specifically to handle the disabilities that the individual has."[53] The dog is then reassessed every two weeks to see if any retraining is required or if the veteran has had any physical deterioration in their mobility or their mindset.

"We had one veteran who we gave a dog to, a young woman who was basically blown up and thrown. She didn't lose any of her arms, her legs, but her body was so traumatized, bones were shattered, this that and the other thing, and because of all of this her immune system wasn't as strong. And she basically had arthritis that you probably would see in an eighty-year-old. So, her dog had to be trained to carry a lot of things for her."

Tower of Hope has since gone on to work to provide service dogs to people who have other disabilities and illnesses, such as autism, multiple sclerosis, muscular dystrophy, spinal injuries, and others. They are proof, like K9s for Warriors, that dogs can help humans back from a difficult past.

Dogs are especially attuned to unconditional love. Marion agrees. She thinks that there is a difference between rescue dogs and dogs in general. "Dogs allow people to establish a relationship where there really is unconditional love. That's one of the things about dogs, that every time you come home, they're so happy to see you. It's as if it's the first time."

She knows that taking care of a dog can be very emotionally rewarding—especially for somebody who has very little control over his or her life—because that dog is dependent on you, so you have unconditional love. "You have a being who is dependent on you, and you're responsible for that being. I think having a sense of responsibility can be very healthy for people who have undergone psychological trauma or who have had certain freedoms taken away for whatever reason." This is one of the reasons she believes that dogs can be essential for teaching children compassion and empathy. They can see how to take care of somebody who is even smaller than they are and yet needs them desperately.

Rebecca from ReCHAI said that dogs can help in smaller but no less significant ways. The humans in the equation are people in transition, not just veterans and prison inmates, but also abused children, families of children who have autism, and older adults.

It is not so much what happens to us in life but how we act to these events. That, my friends, is easier said than manifested. We never saw George's illness coming. We always took George for his regular vet visits. Somehow this serious illness that had been building up over time was missed.

We brought George back to the 24/7 animal hospital and discussed what would be best for George. The doctors had drained the fluids from his body cavity and he seemed like his old happy self, thrilled to see us. There was no trace of any illness. But, the vet cautioned, this would only be temporary. It felt like a horrible deranged fairy tale—George was fine now but it would all come crashing down when the fluids built up again. He looked so healthy and happy—how could it be possible that his life was nearing an end? He was back to being the George we knew.

Paving the Way for True Understanding

In Buddhism there are two kinds of understanding. There is the "understanding" that is really knowledge, the buildup of facts that we have assimilated—knowing accordingly (*anubodha*). It is an understanding that lies more on the surface.

Truly deep understanding or "penetration" (*pativedha*) is perceiving something in its authentic nature. There is no need for names and categorizations or classifications. Meditation can clear the mind's room to allow this deep understanding in.

Meditating is one way to come to terms with how you look at certain events that became a part of your life and who you are. Find a quiet place conducive to thinking and reflection. If your dog is sitting by your side, you can let her peacefulness guide you. Go over what in your life you have found or are finding that you feel you just cannot get over. If what happened has already happened, why attach so much weight to it? It is done and removing it from the present can help pave the way for healing.

That event that occurred in the past is like the skin that a snake has shed. It can look to the casual observer to be the snake itself, with the same shape of a snake. But it isn't. It is just a remnant of the past and the snake has gone off somewhere with a new skin.

CHAPTER 12

THE EIGHTH STEP: THE RIGHT CONCENTRATION

"The quieter you become,
the more you are able to hear."
—Lao Tzu

Is it possible to focus with laser intensity on just one goal or just one thought or just one task? Do we always have to give in to the pull of multitasking? Think about it. Multitasking means doing many things at the same time. Sometimes you have to do several things at once. It's just the way it is. But too many times multitasking means doing many things at once and doing them badly because you simply cannot give your all to more than one important thing at a time. For a while there, multitasking was seen as a badge of honor, proof that you had what it takes to get along in our multi-technology world.

But then it started to become apparent that multitasking wasn't really the timesaver we thought it was. And usually the outcome was a lot less than we expected for our copious efforts.

According to an article in *Psychology Today*, research in the field of neuroscience (the newer multidisciplinary field that encompasses cellular and molecular biology, anatomy and physiology, human behavior and cognition, and other disciplines) has found that our brains are not so good at doing several things at the same time.[54] The brain has to stop and start each time we quickly go from doing one thing to another. We aren't actually doing multiple things at the same time. We just think we are. And we run the real risk of making multiple mistakes and getting worn out.

Better to take the time to do one thing at a time and do each as best as you can. Adopting a dog benefits both the dog and the human. The dog gets (hopefully) a family and reliable food, shelter, and health care. And the human reaps lots of rewards as well. So, if you have one rescued dog and want to multiply the love, what about adopting a second dog? Can you give your love and attention to more than one dog in your household? Since there are so many dogs in shelters, it makes sense to give it some serious thought and to try.

When a Second Dog Enters the Picture

Say you have given this a lot of thought. How to best bring a second dog into your home when you have had a dog living there already? The ASPCA has some good advice about this. Dana Ebbecke said that adopters should first consider their first dog and what kind of playmates he or she likes having around. Once you know your dog's play style and preferences for playmates, this can help narrow down what you should look for in a new companion. Even then it's best to introduce the dogs in a neutral environment.

"Don't force interactions between the dogs, but instead give them time to get comfortable and interact naturally," said Dana. "Once the dogs seem comfortable with each other without appearing fearful or threatening, walk them around the house or building before bringing them inside, and be patient! Sometimes it takes a while to find a good roommate!"

Try to avoid squabbles while the dogs are getting to know each other by removing the toys, chews, food bowls, and any other favorite items of the current dog. Leaving these around can lead to rivalry while the dogs are forming a relationship. "Once the dogs have started to develop a good relationship in the first few weeks," said Dana, "these can be slowly reintroduced. Give each dog his own water and food bowls, bed, and toys. For the first few weeks, only give the dogs toys or chews while they are separated in their crates or other confined areas. Feed the dogs in completely separate areas and pick up the bowls when feeding time is over to avoid competition. Praise your dogs when they are interacting nicely and spend time individually with each dog. If your dogs are very different in age or energy level, be sure to give the older or less energetic dog a private space to rest and enjoy quiet time."

Sherry Woodard of Best Friends offered some thoughts about how to handle adopting a second dog when you already have another dog who has been in the home for a number of years, especially if the first dog has some issues of her or his own.

She said that a trainer or a behavior consultant can help to make sure that the second dog is actually a role model for the dog that has challenges. "Often," she said, "what ends up happening is that the dog with challenges continues to make progress and might even make more progress than it has ever made before because it has a dog to help it. That role model behavior can be a critical key to making more progress that they hadn't seen before."[55]

This is very true. Some dogs have a hard time coping with change, and nothing says change like introducing a new family member into a well-established pack. Charlotte has proven over the years that she likes pretty much all dogs. Taking her to places like dog parks or play groups likely means dealing with dogs of varying temperaments and different states of health, as well as humans with many different ideologies about their dogs, so having play dates with dogs she knows has worked out for her and her dog friends. But what about bringing another pup into her family?

It is not easy to find a dog when there is already a dog well-ensconced in the home. Trying to find the best fit can be daunting. Dog training experts maintain that when introducing a new dog to a home, the humans have to be able to accept that the more recently adopted dog may turn out to be the dominant one. The dogs have to work out this complicated canine hierarchy themselves and nothing the human can do will or should change that. This is not easy to resign yourself to. After all, shouldn't the dog who has been living in the home the longest be the top dog? The answer is no, and it isn't easy to accept this; however, if we are going to accept another dog, we must.

I asked a vet for suggestions about bringing in a second dog into the home. It depends on the dynamic of the household and the personalities of the two dogs. "Generally, respecting the existing pet's territory and existing relationships with the people is helpful," the vet said. But then, as things settle in, it is helpful to carefully observe and evaluate the evolving relationships between the dogs and follow their natural tendencies and trying not to force any particular social construction. "I often observe that it takes months for things to shake out," the vet continued. "I believe that many initial conflicts and issues work themselves out in three to five months. If there is the threat of serious aggression, a behaviorist should be consulted, and separation may be necessary for safety."

This had been brought home to me when someone I knew was considering an older male dog as a friend for their well-ensconced but somewhat lonely female lab mix. She decided that fostering a second dog would be the way to go, to see if both she and her other dog could manage this. She was given an older male dog, Rascal, to foster. With two dogs now in her house, it was hard for my friend to know which one would be the dominant one. After a couple of weeks or so, however, it became clear that who was dominant and who was submissive was not the most important issue.

While at first the two dogs got along and played well enough together, some issues were surfacing that had more to do with people. Rascal was a dog who had been rescued from a hoarding situation and his contact with humans had been extremely limited. While no one knew for sure, it is likely that he had to fend for himself when it came to getting fed. The situation may have been such that whoever dog got the attention of whoever was giving out the food was the dog that actually was able to get the food and eat it. This started to show a bit later in the fostering process. There were other concerns, too.

Taking a dog to a vet is a necessary and important part of the dog's life and something that the dog and the human will need to do on a regular basis if the human is going to be a responsible person. My friend brought Rascal to a couple of different vets and the outcomes were not so good. It became clear after the first visit to a vet that Rascal had some issues that my friend was not aware of. The vet got as far as giving Rascal one shot when the dog reacted like a wild coyote and with a yelping scream, tried to attack the vet. The vet, a congenial and pretty unflappable person, told her that Rascal needed to be muzzled the next time he came in.

While having Rascal wear a muzzle seemed doable to my friend—how hard could it be?—a little bell went off. Just how aggressive was he? But lots of times dogs need to be muzzled when they visit the vet. We all have seen this before. I remember that when we brought George to the vet office, there was one vet who insisted that the dog have a muzzle while other vets would wave off any suggestion that the dog needed to wear a muzzle when being examined by the vet. Whether the dog needed to be muzzled depended on which vet he was seeing. This happens. But with Rascal, this led to another issue. He most definitely did not want to wear a muzzle. But if he was ever going to be seen by a vet, he would have to learn to wear a muzzle.

My friend did her research and found the most comfortable muzzle possible. She put a bit peanut butter inside of the muzzle to make the whole process more agreeable. She would try patiently many times a day to have him sit and get accustomed to the muzzle. Rascal, however, would have none of it. Each time my friend could get no further than having his face sort of be inside the muzzle. But if he was going to need to go to the vet, he was going to need to wear this muzzle.

So, my friend figured she would try a different vet, at a bigger animal hospital. Maybe she would have better luck there. She had seen other dogs who came in not wearing any kind of muzzle but who were wearing a soft muzzle when the vet or the vet tech was bringing the dog back to meet his or her humans. Apparently, some dogs just needed a muzzle when they were involved in something that was helpful for them but was stressful to them.

She made an appointment with a vet there to bring Rascal in for the remainder of his shots and a checkup. She carefully explained on the phone that Rascal had a hard time going to the vet and that he really had never had had a complete examination. This last point had somehow been overlooked when she had agreed to foster him. But they weren't quite prepared, as it turned out. Rascal came back with a vet tech who said they were not able to even examine Rascal. The vet said that sometimes as a new dog gets comfortable in new surroundings, their real personality surfaces.

Looking back, my friend could see that some of the warning signs were there. She just did not see them. I told her about a trainer known for working with difficult dogs and he came to her place to work with Rascal.

Several lessons later, the trainer concluded that Rascal had some longstanding issues with aggression and this primarily involved people. Things were definitely going in the wrong direction with Rascal. As he was getting more comfortable, he was getting more aggressive, and it came to the point where my friend and her family all had a genuine concern for the safety of other dogs and people. My friend would not be adopting Rascal, but she could provide some helpful information about what his needs were. Dogs who have been in hoarding situations, especially for as long as Rascal likely was, can have a range of problems that need to be managed, such as increased attachment issues with humans and not wanting to be touched. He needed someone with the time and means to provide him with complete at-home care—a veterinarian or vet tech on call, and no-expense-spared care, training and everything done for him in the home. She later found out he was adopted into the perfect home for him.

A Puppy Friend?

As it turned out, while Charlotte seemed quite submissive during her first three years, there was a dominate powerhouse brewing in her, and she is definitely a dominant dog. So, bringing into her universe a puppy that had (at his young age) exhibited many traits of a submissive dog seemed like the way to go to bring a friend into her life.

This meant a period of adjustment for Charlotte and for all of the humans involved, too. After all, a puppy is pretty much like a baby, so that meant letting the two dogs get to know each other as well as possible but not permitting full-blown play just yet because of the difference in maturity and size. But Charlotte took to the new puppy, Maverick, almost immediately, and her "mothering" instincts kicked in. She was gentle with him even though he was very energetic and nippy (those puppy teeth bring to mind piranhas) and she got down to his level to gently paw him. It was amazing to see how he learned from her and he responded to her moves and tried to mimic them.

Maverick was adopted by Bonnie, a very good friend, and while Maverick didn't live with Charlotte full time, he was with her a lot and they spent much time getting to know each other.

Sometimes change is difficult and so is conflict, even at the canine level. Watching the two play at times was pretty intense. I had to remember that the dogs can work it out, with my and Bonnie's guidance. Fear that something would happen accidentally because of the difference in the weight of the two dogs had me ready to pull back whichever dog I was holding. Gradually I learned to relax and let them work it out—within reason.

Dogs at play often engage in face mouthing or nipping or biting, and this can seem pretty scary. The sounds they make while having a good time getting to know each other can easily be mistaken for serious growling, but in this case, they were sounds of getting acquainted, and it was amazing to watch this action with an older female dog and a very young male puppy. She was teaching him as if he were a member of her family. He was responding in kind.

It must have been strange for Charlotte. When she was first introduced to the puppy, Charlotte was a little over four years old. Her weight was around forty-eight pounds. The puppy, Maverick, was about sixteen pounds and he was about four months old. This was, as is often the case, quite a bit younger than his adoption paperwork had stated. As the days and weeks went by, though, we could see that the puppy, who had remarkably huge feet, was going to get a lot—a whole lot—bigger.

At first, Maverick was smaller than Charlotte. She would get down to the floor to his level to interact with him. After three months, though, he was starting to get bigger than her. I wonder how she must have felt as the puppy grew bigger almost every time they met. Of course, dogs rely most on their sense of smell, and Maverick was the same Maverick and was still submissive to Charlotte, but he was starting to become bigger than her. Maverick still acted very much like the puppy he was. He was just a big puppy. A very big puppy, who it turned out, had some special needs.

Maverick was adopted from a rescue organization and he was very young at the time of his adoption. He had a health record, but not much of one. He was tested for various ailments that can come with a puppy or dog from unknown origins. He had giardia and another parasite and received medication for these. He seemed easy-going and very friendly, despite the discomfort the giardia bestowed on his digestive system. He would lose weight even though he ate. He drank water as if it was the most precious things he had ever come across. He would drink two cups as if it were nothing and beg for more. Could he have diabetes? No, the vet didn't think so. He was just drinking inappropriately as some puppies do.

At first, no one thought this was odd but then, after several months and many more visits to the vet, things gradually, very gradually, frustratingly gradually, started to become clearer. There was something very odd going on with Maverick. Things started to add up to something that maybe the vet was not seeing.

A few tests were done to see if a cause could be found and waiting for the results was not easy. And as usually happens with anything medical—human or animal—when the medical person calls to talk about the results of any testing, instead of leaving a message saying the equivalent of "Everything is fine," if that is the case, they just leave a message asking to be called back. But when you return the call, it is hard to reach the person who called, or they have left for the day so the dog mom or dog dad starts imagining all sorts of horrible things. And there will be no answers until the next day.

It turned out that Maverick might have an issue with his adrenal glands as he gets older but that he seemed to be doing okay aside from that. He might have to be checked down the road to make sure they are functioning properly.

The short-term relief was tempered with what would have to be long-term vigilance. At least, however, Bonnie now had some warning of what might happen in the future with her beloved pup, but she also had some time to figure a way to budget for the medication he might need at a later date. While she did have pet health insurance for Maverick, it was not always possible to accurately gauge if the insurance company would cover any of the medical expenses Maverick might likely face in the years to come.

But one thing was clear: Bonnie was committed to Maverick, and that is something I understood well. While knowing that Maverick might end up having physical problems when he is older lends an air of uneasiness to the present, this is a reminder that we cannot live in the past or the future. While yes, Maverick is a big, sweet, and lovable puppy and thinking about what might happen—what is likely to happen—to him in a few years, can be fearful, the truth is that no one knows what the future will bring for the pets we love. We can be as vigilant as we can and try to keep them healthy and out of harm's way but, as George and Alice taught me, we never can know what the future has in store for any of us. But for now, Maverick has gotten over the hump and is gaining the weight and stamina he lost.

If you are depressed you are living in the past.
If you are anxious you are living in the future.
If you are at peace you are living in the present.

—Lao-Tzu

Charlotte has her special needs. Maverick has his. Every dog is special, but some dogs need more care and looking after and we love them, so as the humans who love them and are responsible for them, we are there for them. Knowing your dog has special needs does not change how you feel about them. It only increases the love you have for them. That love is always present, as we are always in the present, or striving to be.

Know What Controls Your Actions

Adopt a puppy or an older dog?

Marion has had many experiences with dogs, some more positive than others. Her family had adopted Mickey, who was technically a puppy at seven months old, but must have had some negative experiences already at this young age. "Even though we love our dogs dearly," she said, "they come with so many unknown triggers and just an unknown past. Sometimes I find it very anxiety-provoking to not know." Marion's family has always been committed to training any dog they adopt, and Mickey was no exception. But no one could ever be sure what would trigger a reaction from him. One of her sons walked into the room holding a motorcycle helmet, and Mickey completely freaked out and started barking uncontrollably. Another time, one of her daughter's friends was grabbing her playfully, and Mickey jumped on her. No harm was done but it was indicative of the trauma that can be a part of some dogs.

Mickey also turned out to be very afraid of men. Marion's husband Ben loves dogs and desperately wanted to connect with Mickey, but the dog would shrink back every time Ben tried to pet him. "Fortunately, Ben was not easily daunted. For more than a year he wooed Mickey with offers of food, slowly getting Mickey used to him. Two years later, Mickey now will come up to Ben on his own, though not as enthusiastically as does with me or my daughter."

Past trauma greatly effects some rescue dogs, but it is also why they work so well with people who have experienced trauma. "I think possibly with the right kind of training, that could be really beautiful because you have a dog that's been through trauma and a person who's been through trauma," she added. "I don't think people realize how much training is involved. If you've got a full-time job, and your kids are going to school, it's hard to provide all the training that a traumatized dog needs."

The Puppy Push

This can occur when taking in a dog from a rescue organization or shelter. The rescue organizations are well-meaning and are eager to have the dogs given good homes. It can happen that some issues just are not noticed, overlooked, or ignored if it might mean the dog might miss the opportunity to get a good home. But the downside is that the people adopting a dog and the dog are not a good fit with each other. Sometimes, especially with the adoption of puppies, there is such a rush to be the first person at wherever the adoption event is held that the person looking to adopt a puppy doesn't have time to get to know more about the puppy and if they already have another dog, how their first dog might feel about a particular puppy. There is no time for the two dogs to get to know each other.

The rescue organizations sometimes put pressure on people wanting to adopt a puppy by insisting that the potential adopters, who have filled out a lengthy adoption form to be approved, which is absolutely essential, come right away not only to come and see the puppy, but to take him or her home right away that very day or lose out on adopting a particular puppy or any puppy. I'm not sure this hurry-up-and-get-here-and-get-the-puppy-and take-him-home approach is good for the human or the puppy.

I asked a vet about the best way to approach finding a dog who will fit with your personality and your household. "I guess I would say common sense goes a long way here. I think it's good if you can meet an animal before being committed to taking it home, though it is really hard to judge what a pet will be like in such a situation."

There is usually more time to think things over and to get to know when it comes to older dogs so the humans and the dog to get to know each other and get a better idea if they can live together and thrive. But this is not always the case. If the potential adopters had time to read between the lines of the potential adopted dog's records and ask detailed questions about the dog's behavior and medical needs, the chances are better for a happy outcome for everyone.

Marion understands that this can happen. "A lot of times, the foster parents, at least the agencies we've dealt with, and I think they're very reputable, but the foster homes often have many dogs. Dogs in a pack are not necessarily the same as a dog with either by itself or one other dog."

This comes back to what Marion said about rescue dogs and commitment. "Owners never know the baggage rescue dogs bring with them. Since Mickey was comfortable with me and my daughter, I had assumed he would be great with people in general. And even though Ben won Mickey over, I get nervous if any other men want to pet him because he seems so scared, and a scared dog is an unpredictable dog."

The Right Concentration

Right Concentration is part of the Mental Discipline component the Eightfold Path, along with the already mentioned Right Effort and

Right Mindfulness. Right Concentration means you are focusing all of your thoughts on one thing. It is the polar opposite of multitasking. Even with two dogs who may want to play with each other when you need quiet, there can be time to meditate. It might be a little more challenging with more than one dog, but it can happen.

Find a place that you and your two dogs can be together quietly. There will be times when they are both tired and want to rest and just be with you. You are sitting with two dogs, but you are not multitasking now. You might try to focus on how you feel now that you have two (or more!) furry family members and how this has brought more happiness into the fold. Or just relax and try to not think. Just be where you are and who you are with. Meditation isn't about trying to concentrate. It is about *not* trying to concentrate.

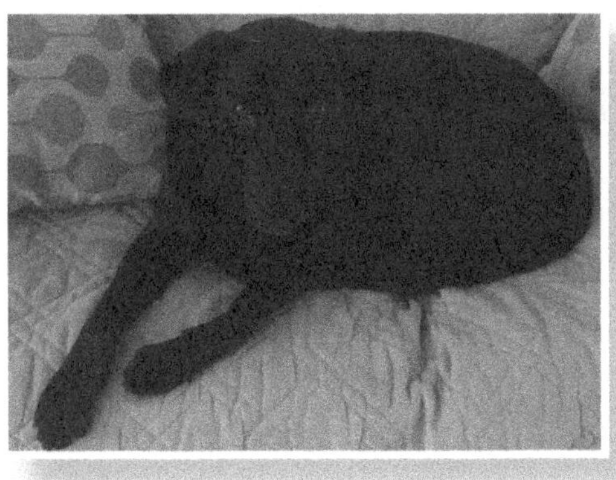

CHAPTER 13

ENLIGHTENMENT OR ONE STEP CLOSER TO ENLIGHTENMENT

We might never reach Enlightenment, but then just trying to reach Enlightenment is reaching it in a way. The striving toward is part of the goal. The journey is the destination. This is what the adopted dog lives. Since she lives in the present, she is always enlightened and is able to be your guide along the path that really never ends. Every dog you adopt will take you at a different pace, over a different time period, on the road to Enlightenment. The journey will always be different. And it will always be enlightening.

Sometimes when you do something, even something you believe to be inherently good, things don't turn out as you had expected. The outcome can be unexpectedly positive or unexpectedly bad. We can never really know 100 percent what the reactions to our actions will be. But we move forward.

When you adopt a rescue dog, you can have the best intentions about welcoming your newest family member, but you never know how your new family member will react to you and his or her new surroundings. It can take longer than expected for the new family member to adapt to new surroundings, new routines, new people, and even new food.

Being proactive and trying to stay ahead of anything that can go wrong doesn't always turn out as expected. Not long after Charlotte's most recent wellness exam, which included bloodwork, concluded she was in good health, she had to go to the ER for a limp she developed after falling while running outside during a play session. The x-ray showed that while the limping likely was the result of a sprain, she also has moderate to severe osteoarthritis in her right coxofemoral joint and moderate osteoarthritis in the left. She is five-and-a-half years old and she has osteoarthritis. Was she born with this? Had she developed it over time?

How to manage this diagnosis with what she loves to do is a challenge, but not an impossible one. We focus on what she can do, not what she cannot do—longer walks and shorter periods of running.

But there was more. The ER doctor noticed something else in Charlotte's health records. She remarked that her urinalysis showed that her urine was dilute. I had heard this before, but no one really had explained what this could mean. It turns out that it means the dog could be starting to develop kidney disease. Another shock. Here I had been buying the best canned food for her and it turns out that it may not have been the best for her after all. Too much protein and too much phosphorous.

Turns out a lot of dog food has excess amounts of phosphorus. To find out what to do I spoke to a veterinarian nutritionist, a growing vet specialty. After a lengthy conversation about Charlotte's health needs and how to reduce the amount of phosphorus and protein in Charlotte's diet, I chose the option to prepare her food at home, using specific recipes formulated by the nutritionist. As a backup, there were an alternative prescription pet food, but a recent recall made that option less likely.

The journey continues, thankfully. More testing will be done to see how she progresses and how we all can work to give her a heathy and happy life.

Maybe it is time to take another look at the Four Noble Truths, and how the first, life is suffering (*Dukkha)*, can be the most difficult to come to terms with. Accepting that there is always going to be suffering in life at first seems counterproductive to having a positive attitude. But it isn't. Since suffering is always going to be there, how we react to suffering is what affects our mindset. Rail against something and you only give it more power. It could take your rescue dog weeks or even months to blend completely into the family fold. Take it day by day and be patient.

The Second Noble Truth (*Samudāya)*, the root of suffering is desire, goes along with being impatient with having the rescue dog feel comfortable with you and everyone in your household as soon as possible. Being frustrated will only fuel anxiety and both of you will lose out.

The Third Noble Truth (*Nirodha)*, the need to stop the desire, takes the above step of recognizing the desire further. It's recognizing that, somehow, we have to stop wanting. So, while you can realize that you want your adopted dog to be one way and he or she is in a different place, making yourself to not want things to be different can lead to both of you being more comfortable. And, this can lead more easily to a happy existence for both of you.

The Fourth Noble Truth (*Magga),* that we are able to stop the desires by following the Eightfold Path, allows us to find a way to stop wanting so much, and to recognize the deep difference between wanting and needing. Really needing. What we really need is usually very different from what we want.

The Eightfold Path is not a long trail that you travel one separate and individual step at a time; it is more of a journey of interwoven stages. To start on the journey of the Eightfold Path we need to have The Right View or Understanding (*Samma ditthi).* This doesn't mean that one way of thinking is inherently wrong while another way of thinking is inherently right. Think about what you are thinking. Have such a deep understanding of the Four Noble Truths that they become a part of your thinking. Recognizing that change is always with us and that nothing is permanent is a way to help with understanding.

Right Intention or Right Thought (*Samma sankappa),* contains many layers and asks us to be truthful about why we are doing something. What we think we become. We take good care of our adopted dogs and want them to be a part of our life. But are we being honest about why we want this? Are we willing to give up something so that our dog can have the essentials? The adopted dog might need extra medical care or extra training. Or both.

Right Speech (*Samma vaca*) urges us to speak kindly of others. Be aware of what you say and know what you are talking about. Generalizations do not apply to individual dogs any more than they apply to individual people.

Right Action (*Samma kammanta*) means that if you have been given gifts, you should use them for the benefit of others. Enjoy the rewards of your hard work and the light that shines on you. But also make time in your life for those who could benefit greatly from your noticing them.

Right Livelihood (*Samma ajiva*) reminds that you must work to earn a living. But in doing so, ensure that no harm comes to others, human or animal.

The Right Effort (*Samma vayama*) shares the wisdom that if there is no effort, there will be no results. Nothing will get done. If, even in some small way, you can work toward helping animals in shelters, the results will be much greater than the smallest effort.

The Right Mindfulness (*Samma sati*) is about knowing you are here now and that this moment is all you have. This is how your dog sees you. He or she watches your signals, even the ones you aren't aware you are giving, and is in sync with your every move.

Right concentration (*Samma samadhi*) says every step leads seamlessly to others.

And this leads to *Dhyana,* which encompasses four stages of meditation, which in turn lead to deep calmness and awareness.

"He who knows others is wise.
He who knows himself is enlightened."
—Laozi

These are just some of the routes of and to Zen. One always leads to another and sometimes in unexpected ways. For a dog, the journey is the goal and you are his or her life. And no matter what happens, he or she will always be in your heart. The more you share your love of rescued dogs, the more love you will have. Always.

Sharing your life with a dog really does keep you grounded and in the present. And, just as one thing leads to another in caring for your dog, the path to Enlightenment involves a natural continuation of steps. One step leads to the next and the last step in one stairway is really the first step in another stairway. The step you take today will lead to the step you take tomorrow, and so on.

Sharing life with a pet who has been rescued takes us to a deeper understanding. We know that our dog has likely had a difficult past. Our dog may still have imprinted memories of this past but he or she is still focused on you. Just as Zen needs to be experienced rather than learned, so it is with a rescue dog. By adopting them and caring for them, we learn more about ourselves. And hopefully, how to be more aware of what is going on in the present for both you and your dog.

Living with a dog can benefit the youngest humans in a household, as well. An article in *Nature*[56] pointed out that after two decades of research, it was very apparent that children who grow up around dogs have lower rates of asthma than those who don't. Many researchers say this is down to something called that hygiene hypothesis—the premise that being exposed to a little dirt early in life can protect against allergic diseases that can emerge later in life. While this idea has long been appealing, until a recent study there was no concrete data to show a relationship between owning a dog and better immune health.

A research team led by Anita Kozyrskyj, a pediatric epidemiologist at the University of Alberta in Edmonton, Canada, discovered that babies who had pets in their home had a better range of microbes in their system than babies who lived in homes with no pets. Even as recently as a generation ago, these research findings would have caused alarm. The thinking was at the time that microbes equaled germs and were therefore something to be avoided at all costs. If there were allergies in a family, the doctor might have even recommended that the family pet be removed.

Now, however, we know that the immune system develops alongside what is going on inside our gut. If infants are growing up with practically no exposure to microbes, the kinds that are a given with a dog's coat or what they bring into a house with their feet, the child's immune system may later decide that these elements should be attacked. Spending a lot of time indoors, which is often the case in developed countries, is also a factor. Dogs can be a way to bring a little dirt safely into an infant's surroundings.[57]

Adopting a Rescue Dog Is a Commitment to Both of You

As Marion, who has adopted several rescue dogs over the years with her family said, "It's taught us all a lot about commitment because rescue dogs can come with baggage." Marion and her husband and two children adopted a terrier whom they named Cindy, who is now about twelve years old. "She was a little dog when we got her. She was our second rescue dog, and our daughter wanted a small dog, so we adopted this adorable small terrier mix. When we first got her, she was so scared and could barely eat. I remember just lying with her on a bed and getting her to feel comfortable."

But then things changed. "Then within two weeks, she was a real terrier, and she was snapping at everybody. I remember my daughter crying at one point when Cindy had snapped at her." Marion's twelve-year-old daughter told her mom that this was not how she thought it would be. So, they discussed her daughter's expectations. "We talked about that, because the agency will always take a dog back. Did we want to give Cindy back because it wasn't what she was expecting?" Marion's daughter was aghast and said no. She didn't want to lose her. So, they worked with Cindy. They trained her and worked hard with her and now the terrier is fine and at home with the family. Things are not perfect, but they never are, and anyone who expects perfection from anything is going to be perpetually disappointed. "Cindy still snaps at strangers unpredictably, but I think that's an important lesson for children, that relationship. I think it's easier for them to explore this with animals sometimes. If you notice, a lot of children's programming is either animals or trains or things that are not human."

But we all have to remember that relationships do have bumps. Marion asked her daughter what she was willing to do. "Are you willing to go to a training class with Mommy to learn how to help Cindy be a better dog?" Her daughter definitely did. "I think it taught us all about commitment."

For You and Your Dog

For this final exercise offered in this book, you might consider a combination of meditation and writing. Think about how your life has changed with the adoption of your dog. Or if you haven't been able to adopt a rescue dog, think about how other rescue dog-related activities—volunteering at and animal shelter, helping to spread the word about homeless pets, whatever you have done—have changed you.

Then, why not write about it?

"Until one has loved an animal, a part of one's soul remains unawakened."
—Anatole France

REFERENCES

Ashwin. 2015. *Why Do Dogs Have Such a Great Sense of Smell?* Accessed December 14, 2017. https://www.scienceabc.com/nature/animals/why-dogs-sense-of-smell-is-so-good.html.

2018. *Association for Animal Welfare Advancement.* Accessed 2018. https://theaawa.org/page/aboutsawa.

2018. Best Friends. Accessed January 16, 2018. https://bestfriends.org/about-best-friends/our-story.

—. 2018. *Celebrity Supporters.* Accessed January 24, 2018. https://bestfriends.org/about/our-partners/celebrity-supporters/emmy-rossum.

—. 2018. *Dog Breed Ban Alternatives.* Accessed February 4, 2018. https://bestfriends.org/resources/pit-bull-terriers/dog-breed-ban-alternatives.

2015. *Best Friends NKLA.* Accessed November 23, 2018. https://nkla.org/about.

2018. *Black Dog Syndrome, Petfinder.* Accessed January 21, 2018. https://www.petfinder.com/pet-adoption/dog-adoption/black-dog-syndrome/.

Brooks, Helen. et al. 2016. *BMC Psychiatry.* May 28. Accessed January 22, 2018. https://bmcpsychiatry.biomedcentral.com/articles/10.1186/s12888-016-1111-3.

Brooks, Helen; Rushton, Kelly; Walker, Sandra; Lovell, Karina; and Rogers, Anne. 2016. *BMC Psychiatry.* May 28. Accessed 2018.

https://bmcpsychiatry.biomedcentral.com/articles/10.1186/s12888-016-1111-3.

Carilli, Cathy, interview by KJ Fallon. 2018. *Founder* (August 16).

2017. *CK9 Sniffing for Solutions.* Accessed December 16, 2017. http://conservationcanines.org/.

Coates, Jennifer. DVM. 2017. *PetMD.* December. Accessed December 21, 2017. https://www.petmd.com/dog/behavior/5-tips-help-pets-deal-grief?roi=echo3-48533362975-45605760-c3fd0e-ae43a94ef1c70a2a2da6dfe769&utm_source=Newsletter&utm_medium=Email&utm_campaign=NWS_12-19-17&utm_content=NWS_PetGrief.

Dale, Steve. 2014. "Dogs have a big-hearted friend in Joan Rivers." *Chicago Tribune.* April 9. Accessed December 19, 2017. https://www.chicagotribune.com/living/ct-xpm-2014-04-09-sns-201404081830-tms-petwrldctnya-a20140409-20140409-story.html.

2018. *Dictionary.com.* Accessed July 2018. https://www.dictionary.com/browse/meditation.

Ebbecke, Dana, interview by KJ Fallon. 2017. *Behavior Counsleor, ASPCA* (December 5).

Flanagin, Jake. 2014. *The Tragedy of America's Dog.* February 28. Accessed February 4, 2018. https://psmag.com/environment/tragedy-americas-dog-pit-bull-75642.

2018. *George W. Bush Presidential Center.* Accessed August 27, 2018. https://www.bushcenter.org/about-the-center/bush-family/bush-family-pets.html.

Gupta, Sunjata. 2017. *Microbiome: Puppy power.* March 30. Accessed January 22, 2018. http://www.nature.com/articles/543S48a.

Haller, Sonja. 2018. *USA Today.* November 18. Accessed November 27, 2018. https://www.usatoday.com/story/life/allthemoms/2018/11/18/joe-biden-adopts-rescue-dog-german-shepherd-named-major/2047503002/.

Hardy, Tom. 2017. *Tom Hardy Dot Org.* June 7. Accessed January 2018, 2018. http://tomhardydotorg.tumblr.com/post/161532868851/i-first-saw-woodstock-running-across-a-turnpike-we.

2017. *Helen Woodward Animal Center.* Accessed January 17, 2018. https://animalcenter.org/about.

Henri, Sean. 2017. *CT Weekender.* February 7. Accessed June 8, 2018. https://ctweekender.com/the-black-dog-of-the-hanging-hills/.

History.com Staff. 2009. *ASPCA Is Founded.* Accessed January 16, 2018. https://www.history.com/this-day-in-history/aspca-is-founded.

Hopler, Whitney. 2017. "Meet Archangel Chamuel, Angel of Peaceful Relationships." *ThoughtCo.* March 7. Accessed September 29, 2018. https://www.thoughtco.com/meet-archangel-chamuel-124076.

Johnson, Rebecca. PhD, RN, FAAN, FNAP. 2017. *Director, ReChai* (November 27). http://rechai.missouri.edu/.

Maddie's Shelter Medicine Program. 2018. *What is Maddie's Fund?* January 13. Accessed January 17, 2018. http://sheltermedicine.vetmed.ufl.edu/about-us/thanks-to-maddie/what-is-maddies-fund/.

Marbach Road Animal Hospital. n.d. *The Incredible Sense of Smell in the Dog.* Accessed December 19, 2017. http://www.savets.org/Pages/DogsIncredibleSenseofSmell.aspx.

Marion, interview by KJ Fallon. 2017. (November 22).

2018. *MuttNation.* Accessed January 24, 2018. https://muttnation.com/foundation/.

Nakano, Craig. 2008. "Black Dog Bias?" *Los Angeles Times.* December 6. Accessed January 4, 2018. http://www.latimes.com/style/la-hm-black6-2008dec06-story.html#axzz2yGcvNTgZ.

Napier, Nancy K. Ph.D. 2014. "The Myth of Multitasking." *Psychology Today.* May 12. Accessed August 2018. https://www.psychologytoday.com/us/blog/creativity-without-borders/201405/the-myth-multitasking.

National Safety Council. 2018. *What Are the Odds of Dying From...* Accessed February 4, 2018. http://www.nsc.org/learn/safety-knowledge/Pages/injury-facts-chart.aspx.

Nett, Randall J. MD1,2, Tracy K. Witte3, PhD, Stacy M. Holzbauer, DVM1,4, Brigid L. Elchos, DVM5, Enzo R. Campagnolo, DVM1,6, Karl J. Musgrave, DVM7, Kris K. Carter, DVM1,8, Katie M. Kurkjian, DVM1,9, Cole Vanicek, DVM10, Daniel R. O'Leary, DVM1,7, Kerry. 2015. *Notes from the Field: Prevalence of Risk Factors for Suicide Among Veterinarians—United States, 2014.* February 13. Accessed Fabruary 3, 2018. https://www.cdc.gov/mmwr/preview/mmwrhtml/mm6405a6.htm.

PetMD. 2018. *Addison's Disease in Dogs.* Accessed January 19, 2018. https://www.petmd.com/dog/conditions/endocrine/c_dg_hypoadrenocorticism.

—. 2017. *Heart Cancer (Hemangiosarcoma) in Dogs.* Accessed December 28, 2017. https://www.petmd.com/dog/conditions/cancer/c_dg_hemangiosarcoma_heart.

2018. *Rachael Ray's Nutrish.* Accessed November 27, 2018. https://nutrish.com/faq/general/rachael-ray-foundation.

2018. *Rachael's Rescue.* Accessed Janaury 25, 2018. https://www.rachaelray.com/rachaels-rescue/.

Rapaport, Lisa. 2017. *Reuters, Can puppies protect babies from allergies and obesity?* April 27. Accessed August 18, 2018. https://www.reuters.com/article/us-health-microbiome-babies-pets/can-puppies-protect-babies-from-allergies-and-obesity-idUSKBN17T319.

2018. *RSPCA Advice and Welfare, Pets, Dogs.* Accessed January 14, 2018. https://www.rspca.org.uk/adviceandwelfare/pets/dogs.

RSPCA. 2018. *Facts and Figures, Key Information.* Accessed January 17, 2018. https://media.rspca.org.uk/media/facts/-/articleName/PressFactsAndFiguresText.

n.d. *RSPCA, Our History.* Accessed January 14, 2018. https://www.rspca.org.uk/whatwedo/whoweare/history.

2018. *Save Our Shelter.* Accessed September 2018. http://saveourshelter.com/.

Seymour, Kristen. 2014. *Is It a Myth That Black Shelter Pets Are Less Likely to Be Adopted?* July 24. Accessed January 21, 2018. http://www.vetstreet.com/our-pet-experts/is-it-a-myth-that-black-shelter-pets-are-less-likely-to-be-adopted.

Simon, Brett, interview by KJ Fallon. 2018. *President, K9 for Warriors* (August 29).

2018. *The Buddhist Centre.* June. Accessed June 2018. https://thebuddhistcentre.com/text/what-meditation.

2018. *Value Penguin.* Accessed August 2018. https://www.valuepenguin.com/life-of-veterinarian.

Walker, Jessica K., Natalie K. Waran, and Clive J. C. Phillips. 2016. *Owners' Perceptions of Their Animal's Behavioural Response to the Loss of an Animal Companion.* November 3. Accessed December 21, 2017. https://www.ncbi.nlm.nih.gov/pmc/articles/PMC5126770/.

Watkins, Tom. 2010. *Paper delves into British veterinarians' high suicide risk.* March 26. Accessed February 3, 2018. http://www.cnn.com/2010/WORLD/europe/03/26/england.veterinarians.suicide/index.html.

Wilson, Sara Logan. 2017. *Canine Dog Journal.* December 21. Accessed February 4, 2018. https://www.caninejournal.com/dog-bite-statistics/.

Woodard, Sherry, interview by KJ Fallon. 2017. *Animal Behavior Consultant* (December 8).

Yuill, Cheryl. DVM, MSc, CVH. 2011. *Wellness Examination in Dogs, VCA Animal Hospitals.* Accessed January 9, 2018. https://vcahospitals.com/know-your-pet/wellness-examination-in-dogs.

ACKNOWLEDGMENTS

The first thank you has to go to my rescue dog for patiently understanding that working on this book meant some time away from her favorite activities. A special thank you goes to my family for their support and patience. Thanks also go to Debra Englander, Madeline Sturgeon, Heather King, Devon Brown, and Kate Post at Post Hill Press. Thanks also go to E. Fallon for editorial support.

Additionally, there are those who provided some background. Margaret Piwonka, DVM MS CVA, Barbara Williamson of Best Friends, Alyssa Fleck of the ASPCA, to name a few of those whose help I appreciate and whose dedication to animals I greatly admire.

ENDNOTES

1 Hopler, Whitney. "Meet Archangel Chamuel, Angel of Peaceful Relationships," ThoughtCo., March 7, 2017. https://www.thoughtco.com/meet-archangel-chamuel-124076.

2 (Dictionary.com 2018)

3 (The Buddhist Centre 2018)

4 (Woodard 2017)

5 Ashwin Vinod, "Why Do Dogs Have Such a Great Sense of Smell?" *ScienceABC*, 2015, accessed December 14, 2017. https://www.scienceabc.com/nature/animals/why-dogs-sense-of-smell-is-so-good.html.

6 "The Incredible Smell of the Dog," Marbach Road Animal Hospital, accessed December 19, 2017.

7 "Sniffing for Solutions," CK9, accessed December 16, 2017. http://conservationcanines.org/.

8 Ebbeke, Dana. Personal interview with KJ Fallon, December 5, 2017.

9 (Woodard 2017)

10 (K. R. Helen Brooks 2016)

11 (Marion 2017)

12 Yuill, Cheryl. "Wellness Examination in Dogs," *VCA Animal Hospitals*, a blog by VCA Animal Hospitals, 2011, accessed

January 9, 2018. https://vcahospitals.com/know-your-pet/wellness-examination-in-dogs.

13 Yuill, "Wellness Examinations."

14 Walker, Jessica K.; Waran, Natalie K.; and Phillips, Clive J.C. "Owners' Perceptions of Their Animal's Behavioural Response to the Loss of an Animal Companion," *Animals (Basel)* 6, no. 11 (November 2016), https://www.ncbi.nlm.nih.gov/pmc/articles/PMC5126770/.

15 Coates, Jennifer. "5 Tips to Help Pets Deal With Grief," PetMD, accessed December 21, 2017. https://www.petmd.com/dog/behavior/5-tips-help-pets-deal-grief?roi=echo3-48533362975-45605760-c3fd0eae43a94ef1c70a2a2da6dfe769&utm_source=Newsletter&utm_medium=Email&utm_campaign=N-WS_12-19-17&utm_content=NWS_PetGrief.

16 Ebbecke, personal interview.

17 Dale, Steve. "Dogs Have a Big-Hearted Friend in Joan Rivers," *Chicago Tribune*, April 9, 2014,. http://articles.chicagotribune.com/2014-04-09/features/sns-201404081830--tms--petwrldct-nya-a20140409-20140409_1_joan-rivers-pekingese-dog.

18 Johnson, Rebecca. Personal interview with KJ Fallon, 2017.

19 Nakano, Craig. "Black Dog Bias?" *Los Angeles Times*, December 6, 2008. http://www.latimes.com/style/la-hm-black6-2008dec06-story.html#axzz2yGcvNTgZ.

20 (Henri 2017)

21 "Black Dog Syndrome," Petfinder, accessed January 21, 2018. https://www.petfinder.com/pet-adoption/dog-adoption/black-dog-syndrome/.

22 Seymour, Kristen. "Is It a Myth that Black Shelter Pets Are Less Likely to Be Adopted?" Vetstreet, July 24, 2018, accessed

January 21, 2018. http://www.vetstreet.com/our-pet-experts/is-it-a-myth-that-black-shelter-pets-are-less-likely-to-be-adopted.

23 Flanagin, Jake. "The Tragedy of America's Dog," *Pacific Standard*, February 28, 2014, accessed February 4, 2018. https://psmag.com/environment/tragedy-americas-dog-pit-bull-75642.

24 Wilson, Sara Logan. "Dog Bite Statistics (How Likely Are You to Get Bit?)," *Canine Journal*, December 21, 2017, accessed February 4, 2018. https://www.caninejournal.com/dog-bite-statistics/.

25 "What Are The Odds of Dying From..." Tools and Resources, National Safety Council, accessed February 4, 2018. http://www.nsc.org/learn/safety-knowledge/Pages/injury-facts-chart.aspx.

26 *Tom Hardy Dot Org*, "June 7, 2017," in *Tom Hardy Dot Org*, a Tumblr blog by Tom Hardy, accessed January 2018. http://tomhardydotorg.tumblr.com/post/161532868851/i-first-saw-woodstock-running-across-a-turnpike-we.

27 "Emmy Rossum," Celebrity Supporters, Best Friends, 2018, accessed January 24, 2018. https://bestfriends.org/about/our-partners/celebrity-supporters/emmy-rossum.

28 "MuttNation," 2018, accessed January 24, 2018. https://muttnation.com/foundation/.

29 "Rachael's Rescue," Rachel Ray, 2018, accessed Janaury 25, 2018. https://www.rachaelray.com/rachaels-rescue/.

30 (Rachael Ray's Nutrish 2018)

31 "Bush Family Pets," About Us, George W. Bush Presidential Center, 2018, accessed August 27, 2018. https://www.bushcenter.org/about-the-center/bush-family/bush-family-pets.html.

32 (Haller 2018)

33 (Save Our Shelter 2018)

34 (Value Penguin 2018)

35 Nett, Randall J. et al., "Notes From the Field: Prevalence of Risk Factors for Suicide Among Veterinarians—United States, 2014," *CDC Morbidity and Mortality Weekly Report (MMWR)* 64, no. 5 (February 13, 2014): 131-132.

36 (Association for Animal Welfare Advancement 2018)

37 Watkins, Tom. "Paper Delves into British Veterinarians' High Suicide Risk," CNN, March 26, 2010, accessed February 3, 2018. http://www.cnn.com/2010/WORLD/europe/03/26/england.veterinarians.suicide/index.html.

38 "Our History," What We Do, RSPCA, 2018, accessed January 14, 2018.

39 "How to Look After and Care for a Dog," Advice and Welfare, RSPCA, 2018, accessed January 14, 2018. https://www.rspca.org.uk/adviceandwelfare/pets/dogs.

40 (RSPCA 2018)

41 History.com Editors, "ASPCA is Founded," HISTORY, last updated August 21, 2018, accessed November 14, 2018. https://www.history.com/this-day-in-history/aspca-is-founded.

42 (History.com Editors, "ASPCA.")

43 "Our Story," Best Friends, 2018, accessed January 16, 2018. https://bestfriends.org/about-best-friends/our-story.

44 "Our Story."

45 (Best Friends NKLA 2015)

46 "What Is Maddie's Fund?" About Us, Maddie's Shelter Medicine Program, January 13, 2018, accessed January 18, 2018. http://sheltermedicine.vetmed.ufl.edu/about-us/thanks-to-maddie/what-is-maddies-fund/.

47 "About," Hellen Woodward Animal Center, 2017, accessed January 17, 2018. https://animalcenter.org/about.

48 "About," Hellen Woodward Animal Center.

49 "About," Hellen Woodward Animal Center.

50 "Heart Cancer (Hemangiosarcoma) in Dogs," PedMD, 2017, accessed December 28, 2017. https://www.petmd.com/dog/conditions/cancer/c_dg_hemangiosarcoma_heart/.

51 (Simon 2018)

52 (Simon 2018)

53 (Carilli 2018)

54 Napier, Nancy K. Ph.D. 2014. Psychology Today. May 12. Accessed August 2018. https://www.psychologytoday.com/us/blog/creativity-without-borders/201405/the-myth-multitasking.

55 (Woodard 2017)

56 Fall, Tove; Ekberg, Sara; Lundholm, Cecilia; Fang, Fang; Almqvist, Catarina. "Dog characteristics and future risk of asthma in children growing up with dogs." *Nature*, November 15, 2018, accessed January 21, 2018. https://www.nature.com/articles/s41598-018-35245-2.

57 Rapaport, Lisa. "Can Puppies Protect Babies from Allergies and Obesity?" *Reuters*, April 27, 2017, accessed August 18, 2018. https://www.reuters.com/article/us-health-microbiome-babies-pets/can-puppies-protect-babies-from-allergies-and-obesity-idUSKBN17T319.

ABOUT THE AUTHOR

KJ Fallon is a journalist and author, and has long been interested in and has studied Zen Buddhism. Over the years, KJ has been rescued by several adopted dogs and it seemed natural to bring the two stories together in the author's second nonfiction book. Pico Iyer, KJ's former colleague and an essayist, novelist, and longtime friend of the Dalai Lama, encouraged KJ to write this book. Pico has been a frequent speaker and interviewer at TED—the nonprofit known for disseminating ideas often with brief, intense discussions.

KJ Fallon is a former reporter with TIME magazine who currently works as a freelance writer for numerous media outlets and as a nonfiction and fiction book editor. KJ has a Master of Arts in applied linguistics from Columbia University.